SpringerBriefs in Astronomy

Series Editors

Martin A. Ratcliffe
Wolfgang Hillebrandt
Michael Inglis

For further volumes:
http://www.springer.com/series/10090

Wolfram Schmidt

Numerical Modelling of Astrophysical Turbulence

 Springer

Wolfram Schmidt
Institut für Astrophysik
Georg-August-Universität
Göttingen
Germany

ISSN 2191-9100   ISSN 2191-9119   (electronic)
ISBN 978-3-319-01474-6   ISBN 978-3-319-01475-3   (eBook)
DOI 10.1007/978-3-319-01475-3
Springer Cham Heidelberg New York Dordrecht London

Library of Congress Control Number: 2013944882

Printed on acid-free paper

Springer is part of Springer Science+Business Media (www.springer.com)

Preface

Turbulence has emerged as an important research topic in several areas of astrophysics, ranging from stellar astrophysics to cosmology. In contrast to terrestrial turbulence, astrophysical turbulence encompasses regimes that are hardly accessible by laboratory measurements. Although pure turbulent dynamics is a scale invariant process, invariance is broken by gravity and thermal processes in astrophysical systems. The situation is further complicated by interactions with magnetic fields, chemical processes, and a diversity of different mechanisms of energy injection. Even for the most idealized systems, supersonic flow introduces severe technical and theoretical difficulties. As a consequence, supercomputer simulations became an essential tool to investigate the properties of astrophysical turbulence.

This book focuses on fundamental statistical properties of hydrodynamical turbulence and numerical methods to perform simulations, from which these properties can be inferred. Computational astrophysics has been particularly active in devising such methods in the last decade. Moreover, the advancement of massively parallel computers made compressible turbulence simulations of sufficient numerical resolution feasible. Thereby, a wealth of data was obtained. Although the interpretation of the data is still not settled, many useful insights have been gained. Basically, there are two schools of thought. One school takes it for granted that statistically stationary and homogeneous turbulence must have universal scaling properties, even in the highly compressible regime. Finding universal statistics is a matter of asking the right question in a numerical experiment. Any deviation from non-universality must be due to resolution effects or other external factors that distort the pure nonlinear dynamics of turbulence. The opposing school of thought emphasizes the complexity of astrophysical turbulence. From this point of view, compressible turbulence is a multiparameter system with at least two parameters (the Mach number and the mixture of solenoidal and compressive large-scale modes). Depending on the system parameters, different statistical properties and scaling laws are observed (both in the literal sense and in the sense of analyzing numerical data). With the material included in this book, I make an attempt to ponder both points of views. Possibly, they convey different aspects of the same thing. I decided to restrict the discussion to the context of supersonic isothermal turbulence in star-forming clouds, which is the simplest form of compressible turbulence. In galaxies such as our Milky Way, stars form in molecular clouds, in which the gas is very cold because it cannot cool further. Consequently, the Mach numbers are high. Since the temperature is close to its minimum, it can be assumed to be nearly constant. The importance of magnetic fields in molecular clouds is undisputed, but the numerical treatment of magnetohydrodynamical turbulence introduces many difficulties, which were fully tackled only at recent time. Notwithstanding their astrophysical shortcomings, simulations of purely hydrodynamical turbulence played a substantial role in fundamental studies of supersonic turbulence. The guided tour through theoretical, numerical, and astrophysical topics in this book considers selected examples. By far not all important works are covered or even mentioned. It is up to reader to draw conclusions and to further follow this endeavor. More realistic numerical models will doubtlessly play a central role in future research. This means that an ever increasing number of astrophysical processes are going to be incorporated with great detail into numerical simulations. Hopefully, this will bring us closer to turbulent flow conditions in astrophysical systems, without obscuring the underlying physics by overly complex models. In any case, numerical techniques have to pay tribute to this development by utilizing adaptive methods for self-gravitating and turbulent gas, subgrid scale models for small-scale physics below the resolution limit, and sophisticated solvers for hydrodynamics and magnetohydrodynamics under extreme conditions.

The research presented in this book covers parts of my Habilitation Thesis at the Universities of Würzburg and Göttingen. I am indebted to Jens Niemeyer, who graciously supported my work and encouraged me to follow new ideas, rather than sticking to opinions. I thank my former Ph.D. Supervisor, Wolfgang Hillebrandt, for initiating this publication. I also express my gratefulness to Christian Klingenberg and Karl Mannheim, who accompanied my habilitation with all kind of advice. Many thanks to the Ph.D. students, postdocs, and collaborators, who helped me with countless contributions. In particular, I thank Christoph Federrath and Ralf Klessen for doing a great job in unraveling the mysteries of turbulence in star-forming clouds and trying to make sense of them. For several years now, Alexei Kritsuk has been someone to make me aware of the subtleties of numerics and turbulence theory. This is something of high value. I acknowledge the permissions from Christoph Federrath and Alexei Kritsuk to reuse figures from their articles. Another important partner for discussions has been Patrick Hennebelle. And I met Dave Collins just at the right time to learn more about the relationship between turbulence and gravity, by using his simulation suite. Without naming them, I finally thank all the other colleagues and friends, whose influence is so important to do good research.

Göttingen, May 2013   Wolfram Schmidt

Contents

Chapter 1
Turbulence Theory

Beginning with the fluid-dynamical equations, theoretical approaches to compressible turbulence are outlined in this chapter. After a brief summary of the main results of the Kolmogorov theory of incompressible turbulence, attempts to treat compressible turbulence with a similar theoretical framework are considered. In the case of isothermal gas, which plays an important role for the theory of star formation, an important aspect is the log-normal distribution of density fluctuations. Scaling relations for a strictly self-similar turbulent cascade are empirically found to break down for higher-order statistics. This is explained by the intermittency of turbulence. A heuristic model that was successfully used to predict scaling properties of incompressible and compressible turbulence is based on a hierarchy of dissipative structure with log-Poisson statistics. This leads to a simple two-parameter model for the relative scaling exponents, which can be determined through experimental measurements or numerical simulations. Apart from the general velocity statistics of compressible turbulence, the density structure of self-gravitating turbulence is an unsolved problem. Crucial questions concern the support of the gas against gravity in the highly non-linear regime and the mass distribution of gravitationally unstable clumps.

1.1 Fundamental Equations of Compressible Fluid Dynamics

The evolution of the mass density ρ, the velocity $\mathbf{v}$ and the specific energy e_{tot} of a neutral fluid is determined by the compressible Navier-Stokes equations [1]:

$$\frac{\mathrm{D}}{\mathrm{D}t}\rho = -\rho\nabla\cdot\mathbf{v}, \tag{1.1}$$

$$\rho\frac{\mathrm{D}}{\mathrm{D}t}\mathbf{v} = \rho(\mathbf{f}+\mathbf{g}) - \nabla P + \nabla\cdot\sigma, \tag{1.2}$$

W. Schmidt, *Numerical Modelling of Astrophysical Turbulence*,
SpringerBriefs in Astronomy, DOI: 10.1007/978-3-319-01475-3_1,
© The Author(s) 2014

$$\rho \frac{D}{Dt} e_{tot} = \rho \mathbf{v} \cdot (\mathbf{f} + \mathbf{g}) - \nabla \cdot (\mathbf{v}P) + \nabla \cdot (\mathbf{v} \cdot \sigma) + \Gamma - \Lambda, \qquad (1.3)$$

where the substantial time derivative is defined by

$$\frac{D}{Dt} = \frac{\partial}{\partial t} + \mathbf{v} \cdot \nabla. \qquad (1.4)$$

For a perfect gas, the total energy per unit mass is given by

$$e_{tot} = e_{int} + e_{kin} = \frac{1}{2} v^2 + \frac{P}{(\gamma - 1)\rho} = \left(\frac{\gamma}{2} \mathcal{M}^2 + \frac{1}{\gamma - 1} \right) \frac{P}{\rho}, \qquad (1.5)$$

where γ is the adiabatic exponent, $\mathcal{M} = v/c_s$ the Mach number, and $c_s = \sqrt{\gamma P/\rho}$ the speed of sound. The pressure P is related to the mass density ρ and the temperature T via the ideal gas law:

$$P = \frac{\rho k_B T}{\mu m_H}. \qquad (1.6)$$

The constants k_B, μ and m_H denote, respectively, the Boltzmann constant, the mean molecular weight and the mass of the hydrogen atom.

Accelerations of the fluid due to mechanical forces and gravity are denoted by $\mathbf{f}$ and $\mathbf{g}$, respectively. An example of an external force $\mathbf{f}$ supplying energy to the fluid is the random driving force that is commonly used in turbulence simulations (see Sect. 2.1). The tensor σ accounts for dissipative processes caused by the viscosity of the fluid. Since dissipation convertes kinetic into internal energy, σ enters the integral from of Eq. (1.3) only as a surface term. Apart from dissipative heating, the internal energy density $\rho e_{int} = P/(\gamma - 1)$ can be altered by radiative processes, as quantified by the heating rate Γ and the cooling rate Λ. In the following, we consider adiabatic gas dynamics, for which $\Gamma = \Lambda = 0$. The thermal conductivity of the gas is also neglected, which is a good approximation for many astrophysical applications.

The gravitational acceleration is given by

$$\mathbf{g} = -\nabla (\phi + \phi_{ext}), \qquad (1.7)$$

where ϕ is the gravitational potential due to the self-gravity of the fluid and the external potential ϕ_{ext} can arise from any other sources, such as stars or dark matter. The potential ϕ is determined by the mass density ρ through the Poisson equation. For vacuum boundary conditions, the Poisson equation takes the simple form

$$\nabla^2 \phi = 4\pi G \rho. \qquad (1.8)$$

For periodic boundary conditions, on the other hand, ϕ is produced by density fluctuations with respect to the global mean $\langle\rho\rangle$ [1]:

$$\nabla^2\phi = 4\pi G(\rho - \langle\rho\rangle). \tag{1.9}$$

Except for hydrostatic equilibrium, the flow generated by gravity changes the mass density $\rho(\mathbf{x}, t)$ and, thus, the potential. For this reason, self-gravity introduces a back-reaction effect.

Energy dissipation arises from the viscosity of the fluid in the presence of velocity gradients. The viscous stress tensor σ is generally defined by [1] [2]

$$\sigma_{ij} = 2\eta \left(S_{ij} - \frac{1}{3}d\delta_{ij} \right) + \zeta d\delta_{ij}, \tag{1.10}$$

where the two coefficients η and ζ are the dynamic and bulk viscosities of the fluid,

$$S_{ij} = \frac{1}{2} \left(\frac{\partial v_i}{\partial x_j} + \frac{\partial v_j}{\partial x_i} \right) \tag{1.11}$$

is the rate of strain and

$$d = \frac{\partial v_i}{\partial x_i} \tag{1.12}$$

the divergence of the flow. The rate of viscous energy dissipation, i. e., the loss of kinetic energy per unit time due to the viscosity of the fluid, is given by

$$\varepsilon = \sigma_{ij} S_{ij} . \tag{1.13}$$

Viscous dissipation is caused by microscopic collisional processes and is indispensable for hydrodynamical turbulence. In the following section, however, we will see that essential properties of turbulence can be described independent of the dissipation mechanism.

1.2 Inertial Range Scaling

In this and the following section, we consider only hydrodynamical turbulence without gravity. In this case, it is possible to obtain exact analytical results for the statistics of stationary isotropic turbulence, which can be tested on numerical data. For recent extensions to magnetohydrodynamical turbulence, see [2, 3].

---

[1] This can be understood as a renormalization of the nominally infinite potential for an infinitely extended mass distribution with positive mean density $\langle\rho\rangle$ to zero for $\rho = \langle\rho\rangle$.

[2] In the fluid dynamics literature, σ_{ij} often includes the pressure term $P\delta_{ij}$. In the formulation used here, it is a seperate term.

1.2.1 Incompressible Turbulence

For given initial and boundary conditions, the solutions of the incompressible Navier-Stokes equations ($d = 0$) can be classified by a single parameter, the Reynolds number

$$\text{Re} = \frac{VL}{\nu}. \tag{1.14}$$

In the above expression, V is a characteristic velocity of the flow and L the largest dynamically relevant length scale. The kinematic viscosity is defined by $\nu = \eta/\rho$. The parameters L and V follow from boundary conditions or are a property of the forces generating the flow. For $\text{Re} \lesssim 1$, the motion of the fluid is dominated by the viscous stresses in Eq. (1.2). As Re becomes larger, the non-linear advection term $(\mathbf{v} \cdot \nabla)\mathbf{v}$ in the substantial time derivative produces increasingly complicated and eventually chaotic flow patterns. Once the flow is completely random, with statistical quantities that are independent of position and direction, the state of *statistically homogeneous and isotropic turbulence* is reached. Altlhough this is of course a mathematical idealization, it is found to describe the properties of flows with high Reynolds numbers rather well. The following empirical laws have been verified in numerous experimental measurements [4]:

1. The random velocity fluctuations on different length scales l can be statistically quantified by the longitudinal structure functions $S_p(l) = \langle \delta v_\parallel^p(l) \rangle$, where $\delta v_\parallel(l) = [\mathbf{v}(\mathbf{x}, t) - \mathbf{v}(\mathbf{x} + \mathbf{l})] \cdot \mathbf{l}/l$ is the velocity difference between spatial positions $\mathbf{x}$ and $\mathbf{x} + \mathbf{l}$ projected along the separation vector $\mathbf{l}$ and the brackets denote the ensemble average. In practical calculations, the ensemble average is approximated by spatial averaging. Of particular importance are the second-order structure functions $S_2(l)$. For homogeneous isotropic turbulence, $S_2(l)$ obeys the power law

$$S_2(l) \sim V^2 \left(\frac{l}{L}\right)^{2/3}. \tag{1.15}$$

2. Statistically stationary turbulence has a constant mean rate of energy dissipation,

$$\langle \varepsilon \rangle \sim \frac{V^3}{L} \tag{1.16}$$

   independent of the viscosity ν.

The law of constant energy dissipation expresses the equilibrium between the rate at which energy is injected into the flow, for example, by mechanical stirring, and the mean rate at which the energy is dissipated into heat. By carrying out a series of equivalent experiments with ever decreasing viscosity (corresponding to increasing Re), one finds that the rate of energy dissipation does not change. This suggests that, in the limit of infinite Reynolds numbers, the statistical properties of turbulence for $l \ll L$ can be expressed purely in terms of $\langle \varepsilon \rangle$. This is basically Kolmogorov's *second universality assumption*, which is also known as *second similarity hypothesis* [5].

In the picture of the *Richardson cascade*, kinetic energy is transferred across a given length scale l by *local* interactions between turbulent eddies. This means that eddies of size smaller, but not much smaller than l drain energy from larger eddies. There are no significant interactions between eddies of largely disparate size. As long as l is much smaller than the energy injection scale L and viscous effects are negligible, the energy flux Π_l, i. e., the mean rate at which energy is transferred from scales larger than l to scales smaller than l, is constant. This defines the *inertial subrange*, for which [4]

$$\Pi_l \sim \frac{v_l^3}{l} \sim \frac{V^3}{L} \sim \langle \varepsilon \rangle. \tag{1.17}$$

Here, $v_l^2 \sim S_2(l)$ is of the order of the mean kinetic energy of eddies on the length scale l and $\tau_l \sim l/v_l$ is the corresponding dynamical time scale so that $\Pi_l \sim v_l^2/\tau_l$. From the constancy of the energy flux and the second similarity hypothesis, it follows that

$$v_l^2 \sim l^{2/3} \varepsilon^{2/3}, \tag{1.18}$$

which is the two-thirds law for the second-order structure function. The spectral formulation of this law is

$$E(k) \propto \varepsilon^{2/3} k^{-5/3}, \tag{1.19}$$

where the energy spectrum function $E(k)$ is defined by integration of the squared velocity modes $\widehat{\mathbf{v}}(k)$ over spherical shells of radius $k = 2\pi/l$ in Fourier space:

$$E(k) = \frac{1}{2} \oint d\Omega_k \, k^2 \, \widehat{\mathbf{v}}(\mathbf{k}) \cdot \widehat{\mathbf{v}}^*(\mathbf{k}). \tag{1.20}$$

A rigorous mathematical derivation under the assumptions of homogeneity, isotropy, and stationarity in the limit of an infinite Reynolds number yields a scaling law for the third-order structure function, the so-called *four-fifths law* [4, 5]:

$$S_3(l) = -\frac{4}{5} \langle \varepsilon \rangle l. \tag{1.21}$$

In contrast to the two-thirds law, the theory also provides the constant of proportionality in this case. By invoking the hypothesis of self-similarity, there must be a unique scaling exponent and, consequently, Eqs. (1.15) and (1.21) can be generalized to the *Kolmogorov-Obukhov law*

$$S_p(l) = C_p \langle \varepsilon \rangle^{p/3} l^{p/3}. \tag{1.22}$$

The above scaling laws are empirically found to be good approximations for nearly homogeneous and isotropic turbulence at large Reynolds numbers. The break-down of the inertial subrange is roughly given by the equality between the eddy turn-over time scale τ_l and the corresponding viscous diffusion time $\sim l^2/\nu$. With Eq. (1.18), we obtain the *Kolmogorv dissipation scale*

$$l_K \sim \left(\frac{\nu^3}{\varepsilon}\right)^{1/4}.$$

(1.23)

As a consequence, the ratio between the integral scale L and l_K is given by [1]

$$\frac{L}{l_K} = \text{Re}^{3/4}.$$

(1.24)

The inertial subrange is constrained by $l_K \ll l \ll L$ and the corresponding number of degrees of freedom $N = (L/l_K)^3 \sim \text{Re}^{9/4}$. Even for turbulent flows with moderate Reynolds numbers of, say, 10^4, N is a very large number. This argument is often used to motivate the difficulty of computing turbulence in direct numerical simulations.

1.2.2 Compressible Turbulence

A heuristic model for the compressible turbulent cascade is proposed in [6]. The simplest generalization of relation (1.17) that accounts for fluctuations of the mass density is a constant *volumetric* energy flux:

$$\Pi_l \sim \frac{\rho_l v_l^3}{l} \sim \langle \varepsilon \rangle,$$

(1.25)

where ρ_l is the mass density fluctuation on the length scale l. This suggests that the mass-weighted velocity variable $\tilde{v} = \rho^{1/3} v$ fulfills a scaling law of the form

$$\tilde{v}_l^p \propto l^{p/3},$$

(1.26)

in analogy to the Kolmogorov-Obukhov law (1.22) for incompressible turbulence. As generalization of the four-fifth law, linear scaling is expected for $\tilde{v}_l^3$ (see also [7, 8]). Although the identification of $\tilde{v}_l^p$ with two-point statistics of compressible turbulence is only tentative, a linear relation for two-point correlation functions is derived from the isothermal compressible Navier-Stokes equations in [9] and numerically verified in [10]. A further implication of this study is that the mean dissipation rate, which appears in the incompressible four-fifth law, has to be replaced by an effective dissipation rate ε_{eff} that accounts for the modulation of the turbulence energy flux by dilatation ($d > 0$) and compression ($d < 0$). For this reason, it is suggested that the relation $\tilde{v}_l^3 \propto \varepsilon_{\text{eff}} l$ is an asymptote of the inertial subrange toward small length scales, while the scaling on larger scale is different. The transition scale is expected to depend on the excitation of turbulence by the large-scale forcing. Although it is difficult to achieve a sufficient range of length scales in numerical simulations to probe both regimes, data from recent simulations of supersonic isothermal turbulence with extremely high resolution appear to support a steeper, non-universal scaling for turbulence produced by strong compressions on large scales [11] (see also Sect. 3.3).

Another rigorous approach to the generalization of the four-fight law to compressible turbulence is made in [12]. They authors derive a general relation for the spatial correlations of densities and fluxes for any quantity that obeys a conservation law with external sources. In the case of a barotropic fluid with pressure $P(\rho)$, the relation

$$\langle [\rho(0)\mathbf{v}(0)] \cdot [\rho(\mathbf{r})\mathbf{v}(\mathbf{r})]v_\|(\mathbf{r}) + \rho(0)v_\|(0)P(\mathbf{r})\rangle = \frac{\bar{\varepsilon}r}{3} \qquad (1.27)$$

is obtained for statistically stationary and isotropic flow. Here, r corresponds to the length scale l, $v_\| = \mathbf{v} \cdot \mathbf{r}/r$ is the longitudinal velocity, and $\bar{\varepsilon} = \langle \rho^2(0)\mathbf{v}(0) \cdot \mathbf{f}(0)\rangle$, where $\mathbf{f}$ is the specific force.[3] As pointed out in [13], the above relation applies only if the force density $\rho\mathbf{f}$ varies on large scales, whereas an energy injection that is confined to the largest scales requires a smooth acceleration $\mathbf{f}$ in simulations of compressible turbulence (see Sect. 2.1). Nevertheless, an approximate relation similar to Eq. (1.27) was verified by numerical data.

Of particular interest is compressible isothermal turbulence, which is commonly used as a model for turbulence in star-forming clouds [14–18]. In this case, $P = c_0^2\rho$, where c_0 is the constant isothermal speed of sound, and the pressure gradient can be expressed as

$$\nabla P = c_0^2\rho\nabla s, \qquad (1.28)$$

where

$$s = \ln\left(\frac{\rho}{\rho_0}\right) \qquad (1.29)$$

is the logarithmic density fluctuation relative to the mean density ρ_0. If viscous stresses are neglected, the gas-dynamical Eqs. (1.1–1.3) in the isothermal limit can be reduced to

$$\frac{D}{Dt}s = d, \qquad (1.30)$$

$$\frac{D}{Dt}\mathbf{v} = -c_0^2\nabla s + \mathbf{g} + \mathbf{f}. \qquad (1.31)$$

The first equation means that the rate of change of s in a fluid element moving with the flow is given by the negative divergence (Eq. 1.12). By assuming that d fluctuates completely randomly, infinitesimal increments $ds_+ = -d_-dt > 0$ (density enhancements) and decrements $ds_- = -d_+dt < 0$ (density reductions) occur with equal probability, independent of the value of s. The central limit theorem then implies a Gaussian distribution of s [19]:

---

[3] In [12], the symbol $\mathbf{f}$ is used for the force density. We keep our definition for consistency in this text.

$$p(s)\,ds = \frac{1}{\sqrt{2\pi\sigma_s^2}}\exp\left[-\frac{(s-\langle s\rangle)^2}{2\sigma_s^2}\right]ds,\qquad(1.32)$$

where the mean logarithmic density fluctuation, $\langle s\rangle$, is related to the standard deviation σ_s by

$$\sigma_s^2 = -2\langle s\rangle.\qquad(1.33)$$

This relation follows from mass conservation:

$$\int_{-\infty}^{\infty} \rho(s)p(s)\,ds = \rho_0 \quad\text{where}\quad \rho(s) = \rho_0\exp(s).\qquad(1.34)$$

The function $p(s)$ is an example for probability density functions (PDF), which play an important role for the statistical analysis of turbulence. Generally, the PDF $p(\mathscr{X})$ of a random variable $\mathscr{X}$ is defined such that

$$P_{\mathscr{X}\le X} = \int_{X_{\min}}^{X} p(\mathscr{X})dX$$

is the cumulative probability to find a value $\le X$.

The PDF (1.32) is called *log-normal*. Approximately log-normal PDFs were found in many numerical simulations of forced supersonic turbulence (see Sect. 2.1). For one-dimensional isothermal gas dynamics without external forces, a log-normal PDF can be easily justified on the basis of the central limit theorem [19], as explained above. It is reasonable that this argument carries over to three-dimensional isotropic turbulence because gas is locally compressed in randomly varying spatial directions. The influence of an external force field, however, is more difficult to pinpoint (see, for example, [7] and [20]). As shown in Sect. 4.1, deviations from the log-normal distribution appear to be possible if turbulence is driven by a large-scale force field.

Physically, the isothermal limit corresponds to the case where the heat produced by the dissipation of kinetic energy is nearly instantaneously removed from the gas by efficient radiative cooling. The notion of energy conservation, such as in the case of adiabatic gas dynamics, is obviously not applicable in this case. The total energy defined by Eq. (1.5) diverges in the limit $\gamma \to 1$, corresponding to $P \propto \rho$. However, it can be shown that the dynamical variable

$$e_{\text{isoth}} = \frac{1}{2}v^2 + c_0^2 s\qquad(1.35)$$

fulfills the conservation law

$$\rho\frac{D}{Dt}e_{\text{isoth}} + \nabla\cdot(\mathbf{v}P) = \rho\mathbf{v}\cdot(\mathbf{f}+\mathbf{g}),\qquad(1.36)$$

analogous to Eq. (1.3) for zero viscosity. It should be noted that energy dissipation in supersonic turbulence can be partially or even entirely caused by shocks. For the kinetic part of the compressible energy spectrum, the definition

$$\mathscr{E}(k) = \frac{1}{4} \oint d\Omega_k \, k^2 \left[\hat{\mathbf{v}}(\mathbf{k}) \cdot \widehat{\rho \mathbf{v}}^*(\mathbf{k}) + \widehat{\rho \mathbf{v}}(\mathbf{k}) \cdot \hat{\mathbf{v}}^*(\mathbf{k}) \right] \tag{1.37}$$

is proposed in [21]. In contrast to energy spectra computed from the squared Fourier modes of $\mathbf{v}$ or $\tilde{\mathbf{v}}$, $\mathscr{E}(k)$ has the correct physical dimension such that

$$\int_0^\infty \mathscr{E}(k) dk = \int \frac{1}{2} \rho v^2 d^3 x = E_{\text{kin}} \tag{1.38}$$

is the total kinetic energy. This is a consequence of Parseval's theorem with appropriate normalization of the Fourier transforms. Since energy is exchanged between kinetic and pressure modes, however, $\mathscr{E}(k)$ does not correspond to an inertial-range spectrum in the sense of the incompressible turbulent cascade. Theoretical approaches such as [8, 9, 12] tackle the problem of resolving this fundamental difference between compressible and incompressible nonlinear fluid dynamics.

1.3 Intermittency

A generalization of the scaling law (1.22) is the refined similarity hypothesis [4, 22]:

$$S_p(l) = C_p l^{p/3} \langle \varepsilon_l^{p/3} \rangle, \tag{1.39}$$

where the local dissipation rate defined by Eq. (1.13) is averaged over regions of size l and the brackets $\langle \, \rangle$ denote the ensemble average. If ε_l itself follows a power law,

$$\langle \varepsilon_l^p \rangle \propto l^{\tau_p}, \tag{1.40}$$

then the scaling exponents of of the p-th order structure function $S_p(l)$ are given by

$$\zeta_p = \frac{p}{3} + \tau_{p/3}. \tag{1.41}$$

Both experiments and numerical simulations indicate that $\tau_{p/3}$ is non-zero and, as a consequence, the exact self-similarity expressed by the Kolmogorov-Obukhov law does not hold for turbulence in nature.

The deviations from a self-similar turbulent cascade are explained by a property of random systems that is called intermittency. As regards turbulence, intermittency means that the cascade is not strictly self-similar because turbulent fluctuations are not space filling or, at a given position in space, occur only during a certain fraction of time. A very simple model of intermittency is the β-model, which assumes a fractal structure. The spatial filling factor of turbulent eddies deceases as a power of the length scale, which is determined by the parameter $0 < \beta < 1$. The limit $\beta = 1$ corresponds to the fully space-filling, self-similar Kolmogorov-Obukhov cascade. Since the β-model makes reasonable predictions only for a limited range of p, it was extended to bi- and multi-fractal models (see [4]).

A different approach is to construct a model for the statistics of the scale-dependent dissipation rate ε_l. For incompressible turbulence, a hierarchy of dissipative structures is proposed in [23], which are characterized by the ratios of the moments of ε_l:

$$\varepsilon_l^{(p)} = \frac{\langle \varepsilon_l^{p+1} \rangle}{\langle \varepsilon_l^p \rangle}. \tag{1.42}$$

The structures that produce the most intense dissipation are associated with $\varepsilon_l^{(\infty)} = \lim_{p \to \infty} \varepsilon_l^{(p)}$. In the limit of infinite Reynolds numbers, these structures are considered to be infinitely thin vortex filaments. They are called the most intermittent or most singular dissipative structures because their scale dependence is divergent for $l \to 0$:

$$\varepsilon_l^{(\infty)} \sim \lim_{p \to \infty} \frac{l^{\tau_{p+1}}}{l^{\tau_p}} \sim \langle \varepsilon \rangle \left(\frac{l}{L} \right)^{-\Delta}. \tag{1.43}$$

This scaling relation follows from the argument that $\varepsilon_l^{(\infty)} \sim V^2/\tau_l$, where V^2 is the largest amount of kinetic energy that can be dissipated and τ_l is assumed to be a universal time scale associated with eddies of size l. For incompressible turbulence, this is just the eddy turn-over time scale $\tau_l \sim \langle \varepsilon \rangle^{-1/3} l^{2/3}$. By substituting $V^2 \sim (\langle \varepsilon \rangle L)^{2/3}$, we obtain Eq. (1.43) with $\Delta = 2/3$ [23]. Thus, the pure Kolmogorov scaling enters here via the anomalous scaling law (1.43) for the strongest dissipative structures. As we shall see in the following, this gives rise to modified scaling exponents with $\tau_p \neq 0$.

It is further assumed that the scalings of different dissipative structures in the hierarchy are connected by the recursive relation

$$\varepsilon_l^{(p+1)} = A_p \varepsilon_l^{(p+1)\beta} \varepsilon_l^{(\infty)1-\beta}, \tag{1.44}$$

with constant coefficients A_p and $0 < \beta < 1$. As in the β-model, the exponent β characterizes the degree of intermittency. For $\beta = 0$, $\varepsilon_l^{(p+1)} \propto \varepsilon_l^{(p)}$ for all l, which implies $\tau_p = 0$. A momentum hierarchy obeying the above recursive relation corresponds to log-Poisson statistics of the dissipation rate [24]. This fact allows us to re-interpret the intermittency model in a more intuitive way. As proposed in [25], the scale-dependent dissipation rates can be related by

$$\varepsilon_{l_2} = W_{l_1 l_2} \varepsilon_{l_1}, \tag{1.45}$$

where the factor $W_{l_1 l_2}$ describes a random multiplicative process:

$$W_{l_1 l_2} = \left(\frac{l_1}{l_2}\right)^{\Delta} \beta^m, \tag{1.46}$$

In the above expression, the first factor accounts for the amplification of dissipative structures, which tends to produce a singular structure (contraction of vortex tubes into a filaments through the stretching mechanism), and the second factor corresponds to a random series of m modulation events, which tend to decrease the dissipation (fragmentation into weaker substructures). If the number of modulations is drawn from a Poisson distribution, it can be shown that the logarithm of the p-th moment of the multiplication factor is given by

$$\log\langle W_{l_1 l_2}^p\rangle = \Delta\left(p - \frac{\beta^p - 1}{\beta - 1}\right)\log\left(\frac{l_1}{l_2}\right) = \tau_p \log\left(\frac{l_1}{l_2}\right). \tag{1.47}$$

The second equality, which is a direct consequence of Eqs. (1.45) and (1.40), implies

$$\tau_p = -\Delta p + \Delta\frac{1 - \beta^p}{1 - \beta}. \tag{1.48}$$

By substituting the above expression for τ_p into Eq. (1.41), we finally obtain the scaling exponents

$$\zeta_p = (1 - \Delta)\frac{p}{3} + \frac{\Delta}{1 - \beta}\left(1 - \beta^{p/3}\right) \tag{1.49}$$

for intermittent turbulence.

Further insight into the meaning of the parameters β and Δ can be gained by considering the asymptotic limit of the exponent τ_p for $p \to \infty$. Let us assume that D is the fractal dimension of the most intermittent dissipative structures. The probability that an arbitrary point in space lies within a distance l from such a structure (and thus belongs to the region over which ε_l is averaged) is $\propto l^{D-3}$. With the anomalous scaling $l^{-\Delta}$ for $p \to \infty$, we expect the average

$$\langle \varepsilon_l^p \rangle \propto l^{-\Delta p + D - 3}.$$

On the other hand, the asymptote

$$\tau_{p+1} \simeq \tau_p + \Delta \quad \Leftrightarrow \quad \tau_p = -\Delta p + C,$$

follows from Eqs. (1.43) and (1.48) in the limit $p \to \infty$. By comparing the two asymptotic expressions for τ_p, it follows that

$$C = D - 3 = \frac{\Delta}{1 - \beta}, \tag{1.50}$$

i. e., C is the *co-dimension* of the most intermittent dissipative structures. In [23], it is suggested that $C = 2$ for one-dimensional filaments. With the Kolmogorov exponent $\Delta = 2/3$, $\beta = 1/3$ is implied. In this case, the scaling exponents are given by the She-Lévêque formula,

$$\zeta_p = \frac{p}{9} + 2\left[1 - \left(\frac{2}{3}\right)^{p/3}\right], \tag{1.51}$$

which was found to reproduce remarkably well many experimental measurements on incompressible turbulent flows.

Since the assumption of incompressibility does not enter the arguments outlined above (apart from $C = 2$), the intermittency model expressed by Eq. (1.49) might also be applicable to turbulence beyond the incompressible limit. However, simulations of turbulence in the supersonic regime show that ζ_3 can become significantly greater than unity (see, for example, [7] and Sect. 3.2.1. This is clearly at odds with Eq. (1.49), which implies $\zeta_3 = 1$ independent of β and Δ. For incompressible turbulence, the significance of ζ_3 lies in the four-fifth law (1.21), which expresses the constancy of the energy flux. But the third-order velocity structure function of supersonic turbulence does not obey linear scaling. This problem can be overcome by applying the relation

$$Z_p = (1 - \Delta)\frac{p}{3} + \frac{\Delta}{1 - \beta}\left(1 - \beta^{p/3}\right) \tag{1.52}$$

for the *relative* scaling exponents $Z_p = \zeta_p/\zeta_3$, which is derived in [24] for arbitrary ζ_3. For strongly supersonic turbulence, where dissipation is mainly caused by two-dimensional shocks, a co-dimension of $C = 1$ can be assumed. If we further assume that the universality of the exponent $\Delta = 2/3$ applies even in the compressible case, it follows that $\beta = 1/3$ and

$$Z_p = \frac{p}{9} + 1 - \left(\frac{1}{3}\right)^{p/3}. \tag{1.53}$$

This is the so-called Kolmogorov-Burgers model [26], for which the Δ is given by the Kolmogorov theory, but the value of β corresponds to shock-dominated (Burgers) turbulence. Some numerical data were indeed found to be consistent with this model [7, 27]. Generally, however, compressible turbulence simulations point toward a class of models with an exponent $\Delta > 2/3$ and a co-dimenension $1 < C < 2$, depending on the Mach number of the flow (see Sect. 3.2.2). Moreover, intermittency corrections for the mass-weighted velocity scalings (Eq. 1.26) are investigated in Sect. 3.3.

1.4 Self-Gravity

Linear stability analysis of the Euler equations [28] for a plane-wave perturbation in gas of constant temperature T_0 and uniform density ρ_0 shows that the perturbation becomes unstable against gravitational collapse if its size exceeds the Jeans length

$$\lambda_J^{(0)} = c_0 \sqrt{\frac{\pi}{G\rho_0}}, \tag{1.54}$$

where $c_0 \propto T_0^{1/2}$ is the isothermal speed of sound. This expression can be generalized to

$$\lambda_J^{(0)} = \frac{a_J^{1/3} c_0}{(G\rho_0)^{1/2}}, \tag{1.55}$$

where a_J is a geometry factor. For the classical Jeans length, $a_J = \pi^{3/2}$. The corresponding critical mass is

$$M_J^{(0)} = \lambda_J \rho_0 = \frac{a_J c_0^3}{G^{3/2} \rho_0^{1/2}}. \tag{1.56}$$

This expression encompasses the Bonnor-Ebert mass, which follows from the virial theorem and specifies the maximal mass of an isothermal gas cloud that can be supported against collapse by its thermal pressure [29, 30]. The Bonnor-Ebert mass M_{BE} is obtained by setting $a_J = 1.18$.

Several attempts have been made to extend the Jeans stability analysis to turbulent gas. Heuristic arguments are customarily based on Chandrasekhar's proposition [31] to substitute c_0 in Eq. (1.54) by an effective speed of sound, which is given by

$$c_{eff,l}^2 = c_0^2 + \frac{1}{3} v_l^2 \tag{1.57}$$

for turbulent velocity fluctuations v_l on the length scale l (also see [32]). The second term on the right-hand side is interpreted as turbulent pressure, but is not clear how it should be combined with the density fluctuations in the case of compressible turbulence.

The most advanced theoretical analysis of self-gravitating turbulence put forward so far makes use of renormalization group theory [33]. Since the method works only for a statistically stationary equilibrium state, one has to assume that the gravitational free-fall time scale is much greater than dynamical time scale of turbulence. This condition will inevitably be violated for the collapse of overdense structures in self-gravitating turbulence. A further limitation is that an effective Jeans length in the fashion of Chandrasekhar can be derived only for length scales above the integral scale of turbulence, but not for turbulent substructure. A non-perturbative theory of the regime in which the free-fall time scale and the dynamical time scale of turbulence are comparable is suggested in [34]. By combining scaling relations from the β-model of turbulence with the assumption of a scale-by-scale energy equipartition between gravity and turbulence, an intermittent hierarchical cloud model is derived.

Similar considerations lead to theoretical predictions of the mass distribution of gravitationally unstable overdense structures in molecular clouds. For brevity, we refer to these structures as clumps.[4] The mass distribution of the clumps, which is called clump mass function, is possibly related to the initial mass function (IMF) of stars [18, 35]. In the remainder of this section, two important analytical theories of the clump mass function are introduced, followed by a general discussion of the fully non-linear dynamics of gravity-driven gas compression.

1.4.1 Analytical Theories of the Clump Mass Function

To specify the mass in self-gravitating structures, mass distributions $\mathcal{N}(M)$ are calculated such that the number of clumps with masses between M and $M + dM$ is $\mathcal{N}(M)dM$. The total number is obtained by integrating over all masses:

$$N_{\text{tot}} = \int_0^\infty \mathcal{N}(M)\, dM. \tag{1.58}$$

The function $\mathcal{N}(M)$, which is called the clump mass function (CMF), can be determined from numerical simulations or analytical theories. In the former case, there are two options. Either a clump-finding algorithm is applied to determine the smallest simply connected structures whose mass exceeds the critical mass for gravitational instability (see Sect. 4.2) or sink particles are used, which collect mass from gravitationally collapsing gas (Sect. 4.3).

Here, we briefly consider two analytical theories of the CMF. The approach by Padoan and Nordlund [36] is based on the following assumptions. Firstly, gravitationally unstable clumps are assumed to be produced by shock compression of nearly isothermal gas. The initial size of the clumps should thus be comparable to the thickness of isothermal post-shock gas. Secondly, the gas is assumed to be fully turbulent, with a power law for the scale-dependence of the velocity fluctuations. Thirdly, the minimal clump mass is given by the thermal Jeans mass (Eq. 1.56). The number of clumps per mass interval is then given by an expression of the form

$$\mathcal{N}(M) \propto M^{-(x+1)} \int_0^M p_{\text{J}}(M')\, dM', \tag{1.59}$$

where the parameter x depends on the scaling exponent of the turbulent velocity fluctuations. The probability density function $p_{\text{J}}(M')$ of the Jeans mass is determined by the PDF of the mass density (see Sect. 1.2.2). The integral specifies the fraction of

---

[4] In other contexts, the term clump may have a more specific or different meaning.

clumps with masses $M' \leq M$ that are unstable according to the Jeans criterion. For clumps of sufficiently high mass, this integral asymptotically approaches a constant. For this reason, the high-mass tail of $\mathcal{N}(M)$ obeys a power law with exponent $-(x + 1)$.

Since the relevant mass range covers several orders of magnitude, it is convenient to replace the linear CMF (1.59) by a logarithmic mass distribution. The linear and logarithmic CMFs are related by

$$\mathcal{N}(M) = \frac{1}{(M_{\mathrm{J}}^{(0)} \tilde{M})} \frac{\mathrm{d}N}{\mathrm{d} \log \tilde{M}}, \tag{1.60}$$

where $M_{\mathrm{J}}^{(0)}$ is the thermal Jeans mass for the mean density ρ_0, and $\tilde{M} := M/M_{\mathrm{J}}^{(0)}$ the normalized mass. The Jeans stability criterion (1.56) implies $\log \tilde{M} = -\frac{1}{2}s$ for the critical mass at density ρ, where $s = \log(\rho/\rho_0)$. By substituting the integration variable, Eq. (1.59) can be written as an integral over the PDF of the logarithmic density fluctuation, $p(s)$:

$$\frac{\mathrm{d}N}{\mathrm{d} \log \tilde{M}} \propto \tilde{M}^{-x} \int_{-2 \log \tilde{M}}^{\infty} p(s) \, \mathrm{d}s. \tag{1.61}$$

Since more mass is needed to collapse at lower density, the upper bound on the clump mass corresponds to a lower bound on the density fluctuation. In [36], the log-normal PDF defined by Eq. (1.32) is assumed.

The CMF (1.61) does not depend on global scales. Obviously, this cannot be physical if $M_{\mathrm{J}}^{(0)} \gtrsim M_{\mathrm{tot}}$, where $M_{\mathrm{tot}} \sim L^3 \rho_0$ is the total mass on the integral scale L of turbulence. For this reason, the Padoan-Nordlund theory has to be considered as an asymptotic description for $M_{\mathrm{J}}^{(0)} \ll M_{\mathrm{tot}}$. Originally, the power-law exponent x was obtained from the jump conditions for shocks in magnetohydrodynamic turbulence, and it was shown that the resulting slope of the high-mass tail is close to the Salpeter slope $x = 1.35$ (see [36]). The slope for isothermal hydrodynamic turbulence, on the other hand, is [37]

$$x = \frac{3}{5 - 2\beta}, \tag{1.62}$$

where β is the exponent of the turbulence energy spectrum (1.20). In the inertial subrange, $E(k) \propto k^{-\beta}$. Alternatively, β can be determined from the second-order structure function via $\beta = 1 + \zeta_2$, where ζ_2 is the inertial-range slope of $S_2(l)$.

The theory of Hennebelle and Chabrier [38] also makes use of power laws for turbulence. In contrast to the Padoan-Nordlund theory, however, the scaling of turbulence does not determine the clump size, but enhances the critical mass of collapse through turbulent pressure support, as expressed by Eq. (1.57). Apart from that, turbulence promotes gravitational collapse by broadening the PDF of the density fluctuations. By utilizing the Press-Schechter statistical formalism for hierarchical structures, they derive a mass distribution that depends on the derivative of $p(s)$ with respect to the length scale R associated with clumps of mass M. For length scales $R \ll L$, this leads to the approximation

$$\mathcal{N}(M) \simeq -\rho_0 \left(M \frac{dM}{dR} \right)^{-1} \frac{ds}{dR} \exp(s) p(s). \tag{1.63}$$

If purely thermal support is assumed, M and R are given by the thermal Jeans mass and length, respectively. By defining the dimensionless scale parameter $\tilde{R} := R/\lambda_J^{(0)}$, where $\lambda_J^{(0)}$ is given by Eq. (1.55), the density-dependence of M and R implies

$$\tilde{M} = \tilde{R} = \exp(-s/2). \tag{1.64}$$

By inverting this relation, it follows from Eq. (1.63) that

$$\frac{dN}{d \log \tilde{M}} = -\frac{\rho_0}{M_J^{(0)}} \tilde{M}^{-3} p(-2 \log \tilde{M}). \tag{1.65}$$

Other than the CMF defined by Eq. (1.61), the above CMF is fully determined by $M_J^{(0)}$ and its high-mass tail does generally not obey a power law.

By assuming that turbulent pressure contributes to the support against gravitational collapse, an implicit relation between the gravitationally unstable mass and the density fluctuation is obtained:

$$\tilde{M} = \tilde{R} \left(1 + \mathcal{M}_*^2 \tilde{R}^{2\eta} \right), \quad s = \log \left(\frac{1 + \mathcal{M}_*^2 \tilde{R}^{2\eta}}{\tilde{R}^2} \right). \tag{1.66}$$

The intensity of turbulence is specified by the Mach number $\mathcal{M}_*$ of turbulent velocity fluctuations on the length scale $\lambda_J^{(0)}$,

$$\mathcal{M}_* := \frac{\mathcal{M}_{\mathrm{rms}}}{\sqrt{3}} \left(\frac{\lambda_J^{(0)}}{L} \right)^\eta, \tag{1.67}$$

where $\eta = (\beta - 1)/2 = \zeta_2/2$ and $\mathcal{M}_{\mathrm{rms}} \sim V/c_0$ for an integral velocity V (a statistical definition of $\mathcal{M}_{\mathrm{rms}}$ as root mean square (RMS) Mach number is given by Eq. 3.1). Since the turbulent pressure on the length scale $\lambda_J^{(0)}$ equals $\mathcal{M}_*^2 \rho c_0^2$, the parameter $\mathcal{M}_*$ measures the relative significance of turbulent versus thermal support for clumps of size $\sim \lambda_J^{(0)}$. The turbulent Mach number on the length scale R is given by $\mathcal{M}_* \tilde{R}^\eta$. The resulting CMF is

$$\frac{dN}{d \log \tilde{M}} = \frac{2\rho_0}{M_J^{(0)}} \frac{1 + (1-\eta) \mathcal{M}_*^2 \tilde{R}^{2\eta}}{\tilde{R}^3 [1 + (2\eta + 1) \mathcal{M}_*^2 \tilde{R}^{2\eta}]} \times p \left[\log(\tilde{M}/\tilde{R}^3) \right]. \tag{1.68}$$

To evaluate the above expression, the dimensionless scale parameter $\tilde{R}$ has to be numerically determined by inversion of Eq. (1.66) for a given value of $\tilde{M}$. A comparison between Eqs. (1.61), (1.65), and (1.68) is made for numerical simulations in Sect. 4.2.

1.4.2 The Rate of Compression

Gravity tends to compress gas, i. e., to produce negative divergence (see Eq. 1.30). Thus, the dynamics of self-gravitating gas can be analyzed by relating the divergence $d = \nabla \cdot \mathbf{v}$ to gravity. By contracting the momentum Eq. (1.2) divided through ρ with the divergence operator $\nabla \cdot$ from the left, a partial differential equation for the time evolution of d is obtained [39, 40]:

$$\frac{Dd}{Dt} = -4\pi G(\rho - \rho_0) + \nabla \cdot \mathbf{f} + \frac{1}{2}\left(\omega^2 - |S|^2\right) - \frac{1}{\rho}\nabla^2 P + \frac{1}{\rho^2}\nabla\rho \cdot \nabla P. \quad (1.69)$$

Here, the Poisson equation (1.9) for periodic boundary conditions was used to substitute $-\nabla \cdot \mathbf{g} = \nabla^2 \phi$ with the source term of the gravitational potential. From the first term, one can see that the vorticity ω contributes to positive divergence, while the rate of strain $|S|^2 = 2S_{ij}S_{ij}$ is associated with negative divergence.

If we define the gas compression by the dimensionless parameter

$$\delta = \frac{\rho}{\rho_0} - 1, \quad (1.70)$$

the above equation can be casted into the form

$$-\frac{Dd}{Dt} = 4\pi G\rho_0\delta - \Lambda - \nabla \cdot \mathbf{f}, \quad (1.71)$$

where $-Dd/Dt$ is the net *rate of compression* experienced by a fluid element moving with flow, $4\pi G\rho_0\delta$ corresponds to the compression rate due to gravity, and

$$\Lambda = \frac{1}{2}\left(\omega^2 - |S|^2\right) - \frac{1}{\rho}\nabla^2 P + \frac{1}{\rho^2}\nabla\rho \cdot \nabla P \quad (1.72)$$

defines the *local support* of the gas against gravity. The contributions of turbulence and thermal gas pressure are

$$\Lambda_{\text{turb}} = \frac{1}{2}\left(\omega^2 - |S|^2\right), \quad (1.73)$$

$$\Lambda_{\text{therm}} = -\frac{1}{\rho}\nabla^2 P + \frac{1}{\rho^2}\nabla\rho \cdot \nabla P. \quad (1.74)$$

For isothermal gas, $P = c_0^2 \rho$ with $c_0 = \text{const.}$, and the thermal support can be expressed in terms of the logarithmic density fluctuation defined by Eq. (1.29):

$$\Lambda_{\text{isoth}} = -c_0^2 \nabla^2 s. \tag{1.75}$$

In the following, it is assumed that direct compression by the specific force $\mathbf{f}$ is negligible compared to Λ_{turb} and Λ_{therm}, which is a reasonable approximation for large-scale forcing.

The classical Jeans length can be obtained from a linear perturbation analysis of the rate of compression (Eq. 1.71) for small density perturbations $\delta \ll 1$. Assuming that the gas is initially at rest, linearization of Eq. (1.1) yields

$$\frac{\partial \delta}{\partial t} \simeq -d',$$

where d' is the divergence corresponding to the density perturbation. With the usual plane-wave ansatz $\delta \propto \exp[i\tilde{\omega}t + \mathbf{k} \cdot \mathbf{x}]$, where $\tilde{\omega}$ is the angular frequency and k the wavenumber, we have $d' = -i\tilde{\omega}\delta$. A dispersion relation is obtained by linearizing Eq. (1.71):

$$\tilde{\omega}^2 \simeq -4\pi G\rho_0 + k^2 c_0^2. \tag{1.76}$$

Hence, the perturbation is unstable if

$$k < k_{\text{J}}^{(0)} = \frac{1}{c_0}\sqrt{4\pi G\rho_0}, \tag{1.77}$$

which is equivalent to Eq. (1.54) for the Jeans length $\lambda_{\text{J}}^{(0)} = 2\pi/k_{\text{J}}^{(0)}$.

In contrast to the above linear perturbation analysis, Eq. (1.71) applies to density variations of any magnitude. In Sect. 4.3, the gravitational compression rate and the support terms are analyzed in the highly non-linear regime of self-gravitating supersonic turbulence. Generally, if $\Lambda > 4\pi G\rho_0\delta > 0$ is sustained for most of the time, the gas is supported against gravity. As δ grows, however, it becomes increasingly improbable that a fluid element can escape the pull of gravity and collapse may ensue. In a collapsing region, $\delta \gg 1$, $4\pi G\rho_0\delta \gg |\Lambda|$, and the local free-fall time scale

$$t_{\text{ff}} \sim (4\pi G\rho_0\delta)^{-1/2}$$

associated with the gravity term becomes the dynamically dominant time scale. In particular, the time scales $1/\omega$ and $1/|S|$ associated with turbulence will be $\lesssim t_{\text{ff}}$. This shows that the underlying assumption of the theoretical analysis in [33] is violated. At the end of collapse, the gas reaches an equilibrium state, in which the volume-averaged support is balanced by gravity, i. e., $\langle \Lambda \rangle \sim \langle 4\pi G\rho_0\delta \rangle$. For an isolated object, this corresponds to virial equilibrium.

References

1. L. Landau, E. Lifshitz, *Fluid Mechanics, Course of Theoretical Physics*, vol. 6, 2nd edn. (Pergamon Press, 1987)
2. A.G. Kritsuk, S.D. Ustyugov, M.L. Norman, P. Padoan, J. Phys. Conf. Ser. **180**(1), 012020 (2009). doi:10.1088/1742-6596/180/1/012020
3. S. Banerjee, S Galtier, Phys. Rev. E **87**(1), (2013). doi:10.1103/PhysRevE.87.013019
4. U. Frisch, *Turbulence. The legacy of A.N. Kolmogorov* (Cambridge University Press, Cambridge, 1995)
5. A.N. Kolmogorov, Dokl. Akad. Nauk SSSR **30**, 301 (1941)
6. M.J. Lighthill, in *Gas Dynamics of Cosmic Clouds, IAU Symposium*, vol. 2 (1955), pp. 1
7. A.G. Kritsuk, M.L. Norman, P. Padoan, R. Wagner, ApJ **665**, 416 (2007). doi:10.1086/519443
8. Aluie, Physica D **247**(1), 54–65 (2013). doi:10.1016/j.physd.2012.12.009
9. S. Galtier, S. Banerjee, Phys. Rev. Lett **107**(13), 134501 (2011). doi:10.1103/PhysRevLett.107.134501
10. Kritsuk, Wagner, Norman, J. Fluid Mech. **729**, R1 (2013). doi:10.1017/jfm.2013.342
11. C. Federrath, On the Universality of Compressible Supersonic Turbulence. Submitted to MNRAS (2013)
12. G. Falkovich, I. Fouxon, Y. Oz, J. Fluid Mech **644**, 465 (2010). doi:10.1017/S0022112009993429
13. R. Wagner, G. Falkovich, A.G. Kritsuk, M.L. Norman, J. Fluid Mech. **713**, 482 (2012). doi:10.1017/jfm.2012.470
14. B.G. Elmegreen, J. Scalo, ARA&A **42**, 211 (2004). doi:10.1146/annurev.astro.41.011802.094859
15. M. Mac Low, R.S. Klessen, Rev. Mod. Phys. **76**, 125 (2004). doi:10.1103/RevModPhys.76.125
16. J. Ballesteros-Paredes, R.S. Klessen, M.M. Mac Low, E. Vazquez-Semadeni, in *Protostars and Planets V*, ed. by B. Reipurth, D. Jewitt, K. Keil (2007), pp. 63–80
17. C.F. McKee, E.C. Ostriker, ARA&A **45**, 565 (2007). doi:10.1146/annurev.astro.45.051806.110602
18. P. Hennebelle, E. Falgarone, A&A Rev. **20**, 55 (2012). doi:10.1007/s00159-012-0055-y
19. T. Passot, E. Vázquez-Semadeni, Phys. Rev. E **58**, 4501 (1998)
20. C. Federrath, R.S. Klessen, W. Schmidt, ApJ **688**, L79 (2008). doi:10.1086/595280
21. J. Pietarila Graham, R. Cameron, M. Schüssler, ApJ **714**, 1606 (2010). doi:10.1088/0004-637X/714/2/1606
22. A.N. Kolmogorov, J. Fluid Mech. **13**, 82 (1962)
23. Z.S. She, E. Leveque, Phys. Rev. Lett. **72**, 336 (1994). doi:10.1103/PhysRevLett.72.336
24. B. Dubrulle, Phys. Rev. Lett. **73**, 959 (1994). doi:10.1103/PhysRevLett.73.959
25. Z.S. She, E.C. Waymire, Phys. Rev. Lett. **74**, 262 (1995). doi:10.1103/PhysRevLett.74.262
26. S. Boldyrev, ApJ **569**, 841 (2002)
27. S. Boldyrev, A. Nordlund, P. Padoan, ApJ **573**, 678 (2002)
28. J.H. Jeans, R. Soci. Lond. Philos. Trans. Ser. A **199**, 1 (1902)
29. W.B. Bonnor, MNRAS **116**, 351 (1956)
30. R. Ebert, Zeitschrift für Astrophysik **37**, 217 (1955)
31. S. Chandrasekhar, R. Soc. London Proc. Ser. A **210**, 26 (1951)
32. S. Bonazzola, J. Heyvaerts, E. Falgarone, M. Perault, J.L. Puget, A&A **172**, 293 (1987)
33. S. Bonazzola, M. Perault, J.L. Puget, J. Heyvaerts, E. Falgarone, J.F. Panis, J. Fluid Mech **245**, 1 (1992). doi:10.1017/S0022112092000326
34. H. Biglari, P.H. Diamond, Physica D Nonlinear Phenomena **37**, 206 (1989). doi:10.1016/0167-2789(89)90130-9
35. P.C. Clark, R.S. Klessen, I.A. Bonnell, MNRAS **379**, 57 (2007). doi:10.1111/j.1365-2966.2007.11896.x
36. P. Padoan, Å. Nordlund, ApJ **576**, 870 (2002). doi:10.1086/341790

37. P. Padoan, Å. Nordlund, A.G. Kritsuk, M.L. Norman, P.S. Li, ApJ **661**, 972 (2007). doi:10.
    1086/516623
38. P. Hennebelle, G. Chabrier, ApJ **684**, 395 (2008). doi:10.1086/589916
39. W. Schmidt, in *Structure Formation, in the Universe*, ed. by G. Chabrier. Cambridge Contem-
    porary Astrophysics (Cambridge University Press, Cambridge, 2009), pp. 20–3
40. W. Schmidt, D.C. Collins, A.G. Kritsuk, MNRAS **431**, 3196–3215 (2013). doi:10.1093/mnras/
    stt399

Chapter 2
Simulation Techniques

An often used method to produced turbulence in numerical simulations is random
forcing. Although this method is a mathematical idealization, it allows us to perform
numerical experiments in the sense that a statistically stationary and isotropic turbulent state is prepared, with well defined statistical properties that can be compared
to analytical theories. For more realistic astrophysical applications, adaptive methods are indispensable. Adaptive mesh refinement was introduced to efficiently treat
flows with inhomogeneous structure. The basic idea is that the numerical resolution is
dynamically adjusted such that subregions of the flow with strong fluctuations or steep
gradients are well resolved, while smoother regions are covered by relatively coarse
grids. The treatment of turbulent flows with adaptive mesh refinement, however, is
still a challenging problem, mainly because refinement criteria that are sensitive to
turbulent eddies or shocks in supersonic turbulence are difficult to formulate. The
need to fully resolve turbulence, however, might be ameliorated by subgrid scale
models, which approximate the effect of numerically unresolved turbulent eddies
and shocks through turbulent viscosity and pressure in large eddy simulations.

2.1 Turbulence Forcing

There is a long record of turbulence simulations in astrophysics, where a force density
$\rho\mathbf{f}$ in Eqs. (1.2) and (1.3) smoothly accelerates the fluid on large scales (in the context
of the ISM, examples are [1–12]). One method is to use a spatially random static
pattern, for which the amplitude is adjusted for each numerical time step so that
power of the forcing is approximately constant [13, 14]. To produce statistically
homogeneous and isotropic turbulence, it is advantageous to compose the force from
randomly evolving Fourier modes. Each mode is given by a stochastic process of a
particular type, the so-called *diffusion process*. A random variable $\mathscr{U}_t$ that evolves as
a diffusion process in time is determined by a drift coefficient $a(\mathscr{U}_t, t)$ and a diffusion
coefficient $b(\mathscr{U}_t, t)$. A fundamental diffusion process with Gaussian statistics is the

W. Schmidt, *Numerical Modelling of Astrophysical Turbulence*,
SpringerBriefs in Astronomy, DOI: 10.1007/978-3-319-01475-3_2,
© The Author(s) 2014

Wiener process, for which $a(\mathcal{U}_t, t) = 0$ and $b(\mathcal{U}_t, t) = 1$. The time evolution of this process is defined by infinitesimal increments

$$d\mathcal{W}_t = \mathcal{W}_{t+dt} - \mathcal{W}_t = \mathcal{N}(0, dt), \tag{2.1}$$

where $\mathcal{N}(0, dt)$ is a normal distribution with mean zero and standard deviation equal to the time increment dt. From the Wiener process, a statistically stationary diffusion process can be constructed by defining

$$a(\mathcal{U}_t, t) = -\frac{\mathcal{U}_t}{t}, \quad b^2(\mathcal{U}_t, t) = \frac{2\sigma^2}{T}. \tag{2.2}$$

This results in a Langevin-type stochastic differential equation:

$$d\mathcal{U}_t = a(\mathcal{U}_t, t)dt + b(\mathcal{U}_t, t)d\mathcal{W}_t = -\mathcal{U}_t \frac{dt}{T} + \left(\frac{2\sigma^2}{T}\right)^{1/2} d\mathcal{W}_t. \tag{2.3}$$

The stochastic process $\mathcal{U}_t$ given by this equation is known as *Ornstein-Uhlenbeck process*.

The mean of $\mathcal{U}_t$ is given by $\langle \mathcal{U}_t \rangle = \langle \mathcal{U}_0 \rangle e^{-t/T}$. Therefore, any information about the initial configuration is exponentially damped over the autocorrelation time scale T. The variance is determined by the differential equation

$$\frac{d}{dt}\langle \mathcal{U}_t^2 \rangle = \frac{2}{T}\left(-\langle \mathcal{U}_t^2 \rangle + \sigma^2\right), \tag{2.4}$$

which has the solution

$$\langle \mathcal{U}_t^2 \rangle = \sigma^2 + \left(\langle \mathcal{U}_0^2 \rangle - \sigma^2\right)e^{-2t/T}. \tag{2.5}$$

If $\langle \mathcal{U}_0^2 \rangle = \sigma^2$, then $\langle \mathcal{U}_t^2 \rangle = \sigma^2$ for all t. Otherwise, the deviation of $\langle \mathcal{U}_t^2 \rangle$ from σ^2 is exponentially damped. Hence, the Ornstein-Uhlenbeck process is *asymptotically stationary* with $\langle \mathcal{U}_\infty^2 \rangle = \sigma^2$. Numerically, the process is readily implemented via the conditional distribution function

$$F(\mathcal{U}_2; t_2 | \mathcal{U}_1; t_1) = \mathcal{N}\left(\mathcal{U}_1 e^{-(t_2-t_1)/T}, \sigma^2\left[1 - e^{-2(t_2-t_1)/T}\right]\right). \tag{2.6}$$

Now, we can define the Fourier modes $\hat{\mathbf{f}}(\mathbf{k}, t)$ of the acceleration field in terms of the Ornstein-Uhlenbeck process [8, 15, 16]:

$$d\hat{\mathbf{f}}(\mathbf{k}, t) = g_\zeta \left[-\hat{\mathbf{f}}(\mathbf{k}, t)\frac{dt}{T} + \frac{V}{T}\left(\frac{2\sigma^2(\mathbf{k})}{T}\right)^{1/2} \mathbf{P}_\zeta(\mathbf{k}) \cdot d\mathcal{W}_t\right], \tag{2.7}$$

where it is understood that $\mathcal{W}_i$ is a vector of three statistically independent Wiener processes. Of course, we also assume that all Fourier modes are given by independent diffusion processes. The normalization coefficient g_ζ and the projection operator $\mathbf{P}_\zeta(\mathbf{k})$ are defined below. The autocorrelation time scale T of the forcing is identified with the large-eddy turn-over time $T = L/V$, where V is the characteristic velocity of the flow, and the integral length $L = X/\alpha$ is set to a fraction $1/\alpha$ of the domain size X. The stochastic diffusion term in Eq. (2.7) ensures that the resulting force field becomes statistically isotropic in physical space, while the drift term causes any anisotropic initial condition to dacay exponentially. This is not guaranteed with the method of steady random forcing, where isotropy must be prepared to very high precision in the initial velocity field.

Ideally, one would choose $\alpha \ll 1$ in order to minimize the effect of periodic boundary conditions [17]. However, this would constrain the dynamical range by far too much for a grid resolution that is computationally feasible. For this reason, $\alpha = 2$ is a typical choice. The wavelength of the force field is then about half of the domain size. Since the isotropy of the flow produced by the forcing increases with the degree of randomness, a relatively large number of modes should be chosen. On the other hand, higher wave numbers must not be contaminated by forcing modes. For this reason, stochastic Fourier modes are usually defined in a narrow window of wavenumbers centered around the characteristic wavenumber $k_0 = 2\pi/L = 2\pi\alpha/X$. In [8], for instance, the variance in Eq. (2.7) is given by

$$\sigma(\mathbf{k}) \propto \begin{cases} k^2(2k_0 - k)^2 & \text{if } 0 < k < 2k_0, \\ 0 & \text{if } k \geq 2k_0. \end{cases} \qquad (2.8)$$

The parabolic form of $\sigma(\mathbf{k})$ corresponds to the energy spectrum $E(k) \propto k^4$ at small wave numbers, which is a common assumption for the energy-containing range.

The mix of transversal (perpendicular to $\mathbf{k}$) and longitudinal (parallel to $\mathbf{k}$) components can be adjusted by means of the projection operator

$$(P_{ij})_\zeta(\mathbf{k}) = \zeta P_{ij}^\perp(\mathbf{k}) + (1 - \zeta)P_{ij}^\parallel(\mathbf{k}) = \zeta\delta_{ij} + (1 - 2\zeta)\frac{k_i k_j}{k^2}. \qquad (2.9)$$

The parameter ζ determines the corresponding Helmholtz decomposition of the force field in physical space. If $\zeta = 1$, all modes are perpendicular to the wavevectors and the resulting force field is purely *solenoidal* (divergence-free). Solenoidal forcing (also called stirring) produces large-scale eddies. In the case $\zeta = 0$, on the other hand, projections parallel to the wave vectors results in a *dilatational* (rotation-free) field. The forcing is commonly called *compressive* in this case, although it produces both compressions ($d < 0$) and rarefactions ($d > 0$).

It is convenient to define the normalization coefficient g_ζ by

$$g_\zeta = \frac{3}{\sqrt{1 - 2\zeta + 3\zeta^2}}. \qquad (2.10)$$

By using the properties of the Ornstein-Uhlenbeck process and definition (2.9), it can be shown that the root mean square (RMS) magnitude of the force in the statistically stationary regime is

$$\langle \hat{f}^2(\mathbf{k}, t)\rangle^{1/2} = \frac{3V}{T}\sigma(\mathbf{k}).$$

With an appropriate normalization of $\sigma(\mathbf{k})$, the physical force following from the inverse Fourier transform has a magnitude $\sim 3V/T$. Once turbulence reaches a steady state, the RMS velocity is about $v_{rms} = \langle v^2\rangle^{1/2} \sim \sqrt{3}V$. The actual flow velocities reached in simulations tend to be less, however, because of the limited efficiency of the energy injection process, particularly if ζ is small (mainly compressive forcing).

As an illustration of the turbulent fields produced by stochastic forcing in numerical simulations with the FLASH code, projections of the density ρ, vorticity $\omega = |\nabla \times \mathbf{v}|$, and divergence $d = \nabla \cdot \mathbf{v}$ onto a plance are shown in Fig. 2.1 (see Chap. 3 and [9] for further details). In one simulation, purely solenoidal forcing ($\zeta = 1$) is applied, and in the other the forcing is compressive ($\zeta = 0$). The RMS Mach number is roughly 5 in both cases. One can see that compressive forcing produces pronounced large-scale structures with higher density contrasts in comparison to solenoidal forcing. As indicated by the vorticity, turbulence tends to be concentrated in the overdense gas in the case of compressive forcing, while vortex filaments are more uniformly distributed if the forcing is solenoidal. The shock fronts produced in the gas are associated with strongly negative divergence.

While solenoidal forcing has been used as the archetypical form of forcing ("turbulence stirring") in many simulations, mixed forcing was introduced in [16]. Supersonic turbulence simulations with dominant compressive components of the forcing [8, 9] were motivated by production mechanisms of turbulence in the interstellar medium, such as gravity-driven flows or supernova shock waves. A theoretical argument is invoked in [4]: Since compressions have a preferred direction, while eddies are two-dimensional structures, a mixture of 1/3 compressive (longitudinal) and 2/3 solenoidal (transversal) modes would be optimal (as shown in [9], this corresponds to $\zeta = 0.5$). Furthermore, it is argued in [18] that solenoidal and compressive modes should be excited in a particular ratio for a given Mach number, otherwise the turbulent cascade will be distorted by the large-scale forcing over a wide range of wavenumbers. Either viewpoint has its merits, depending one what the forcing is supposed to model. If the forcing is a crude model for the energy injection at the largest scales of the system, any mixture could be justified. In this case, however, forcing from the boundaries (e.g. the colliding-flow scenario in [19–22]) might arguably be a more realistic model. On the other hand, if the forcing mimics energy transfer from length scales greater then the computational domain, i.e., the box statistically represents a "piece" of a much larger turbulent flow, then it is plausible that the solenoidal-to-compressive ratio is fixed.

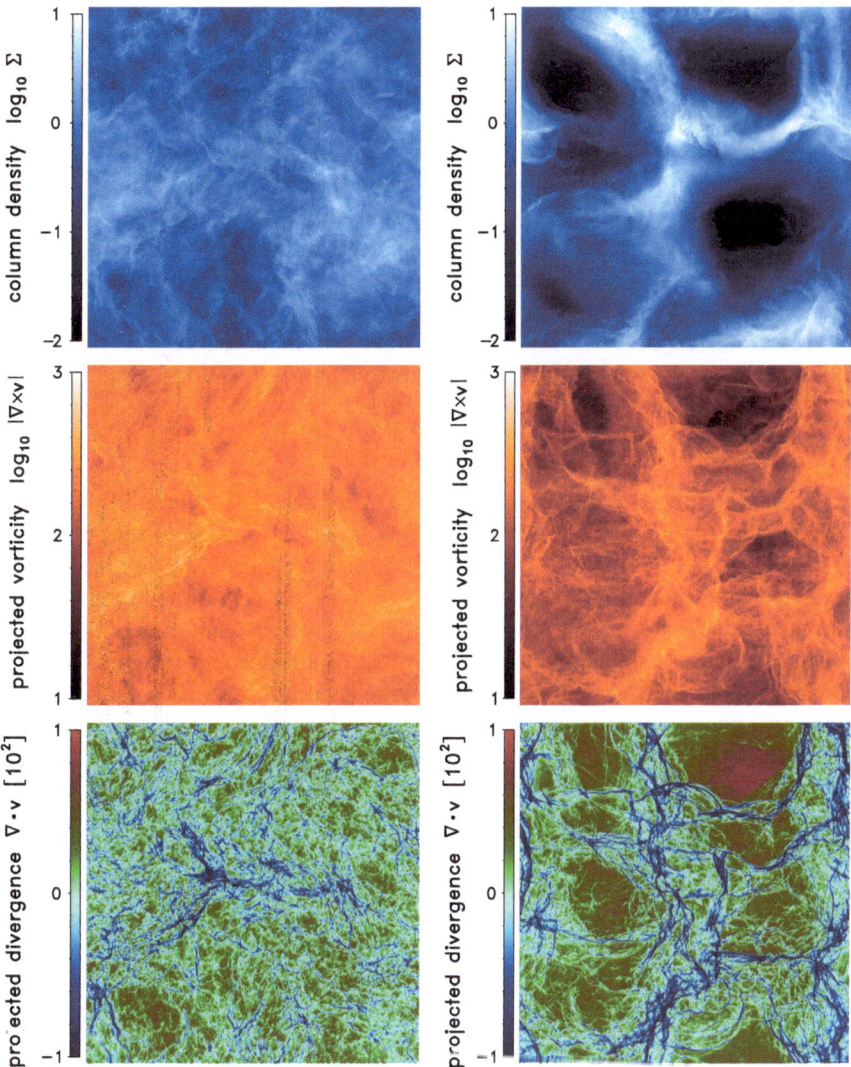

Fig. 2.1 Projections of the gas density (*top*), vorticity (*middle*) and divergence (*bottom*) at time $t = 2\,T$ in FLASH simulations with solenoidal forcing (*left*) and compressive forcing (*right*) [9]. By courtesy of Christoph Federrath

2.2 Adaptive Mesh Refinement

The basic idea of adaptive mesh refinement (AMR) is to dynamically increase the numerical resolution in regions of interest in the course of a simulation. *Block-structured* AMR works with a hierarchy of rectangular grid patches or blocks, which can be created or destroyed according to some refinement criteria [23, 24].

This method is implemented, for example, in the fluid dynamics codes ENZO [25], FLASH [26], and NYX [27]. Usually, grids patches with higher resolution than the parent grids are inserted if a given overdensity is reached or gradients of dynamical variables such as the velocity exceed given thresholds. *Tree-based* AMR codes, on the other hand, create refined regions on a cell-by-cell basis, where the cells are organized in tree structures. An example is the RAMSES code [28, 29].

While AMR is widely used for simulations of self-gravitating systems, the application to turbulence is regarded as infeasible if the turbulent structures fill most of the computational domain. Owing to the intermittency of turbulence, however, turbulent eddies or shocks at a particular instant of time occupy a volume fraction that decreases with length scale. As pointed out in [30], the number of degrees of freedom following from the β-model (see Sect. 1.3) is roughly

$$N \sim \mathrm{Re}^{3D/(D+1)}, \tag{2.12}$$

where D is interpreted as the fractal dimension of dissipative structures. If $1 < D < 2$ for compressible turbulence, then $\mathrm{Re}^{3/2} \lesssim N \lesssim \mathrm{Re}^2$. Compared to $\mathrm{Re}^{9/4}$, which follows from the estimate (1.24) for space-filling turbulence ($D = 3$), the number of degrees of freedom is significantly reduced. It is still very difficult, however, to turn this into an advantage if the limited efficiency to track complicated dynamical structures and the computational overhead of AMR are taken into account.

An essential requirement for computing turbulent flow with AMR is the sensitivity of the refinement criteria on small-scale properties of the flow. For this reason, gradients of the velocity, the density or other fields are used as so-called control variables. The simplest option is to prescribe fixed thresholds for the gradients. It is generally difficult, however, to choose thresholds such that the refinement is efficient (small filling factor of refined regions), while it is guaranteed that all turbulent structures are tracked down (high filling factor). An example are the AMR simulations of galaxy clusters in [31]: To increase the resolution in the vicinity of shocks, refinement is triggered if the difference between the velocities in adjacent cells becomes greater than a given fraction of the local velocity magnitude. However, this introduces a dependence on the motion of the frame of reference and the spatial orientation of the coordinate axes. This can be avoided by using *structural invariants*, i.e., scalars derived from the velocity gradient. A natural choice to track down turbulent eddies is the vorticity modulus ω:

$$q_1 = \frac{1}{2}\omega^2 = \frac{1}{2}|\nabla \times \mathbf{v}|^2 \tag{2.13}$$

Since the steepening of density gradients in the vicinity of shocks is associated with rapidly increasing gas compression, the rate of compression is proposed as additional control variable in [8]. For isothermal gas (see Eqs. 1.69 and 1.75),

$$q_2 = -\frac{Dd}{Dt} = \frac{1}{2}\left(|S|^2 - \omega^2\right) + c_s^2\nabla^2 s \tag{2.14}$$

if the gravity and forcing terms are neglected.

To avoid tunable parameters, statistical moments of the control variables are used in [8] to define the thresholds for refinement. Since the typical magnitude of fluctuations is given by the standard deviation std $q_i(t)$, where

$$\text{std}^2 q_i(t) = \langle q_i^2(\mathbf{x}, t)\rangle - \langle q_i(\mathbf{x}, t)\rangle^2, \tag{2.15}$$

a grid cell is flagged for refinement if any of the control variable deviates by more than one standard deviation from its mean value. Thus, the refinement criterion is

$$q_i' = q_i - \langle q_i(\mathbf{x}, t)\rangle \geq \text{thresh } q_i(t), \tag{2.16}$$

where the threshold is defined by

$$\text{thresh } q_i(t) := \max\left[\langle q_i(\mathbf{x}, t)\rangle, \text{std } q_i(t)\right]. \tag{2.17}$$

This definition ensures that refinement is only triggered by pronounced fluctuations. For a nearly uniform field with $q_i'(\mathbf{x}, t) \ll \langle q_i(\mathbf{x}, t)\rangle$, refinement is inhibited. Only if the mean value is small, for example, in the initial phase, an absolute lower bound has to be specified. For strongly fluctuating fields, the refinement criterion effectively becomes $q_i'(\mathbf{x}, t) \geq \text{std } q_i(t)$. Except for the root grid, for which global averages are calculated, the averaging is constrained to the refined regions at higher levels.

As an example, we consider an AMR simulation of supersonic isothermal turbulence performed with ENZO (see Chap. 3 and [8] for further details). The initial grid (also called the *root grid*) has a resolution of $N_0 = 196^3$. Based on the refinement criteria defined above, refined grids with four times smaller cells are inserted, which results in the effective resolution $N_{1,\text{eff}} = 768^3$. Turbulence is produced by stochastic forcing with $\zeta = 0.1$, as described in Sect. 2.1. Slices of the mass density and the local Mach number of the flow in this simulation are plotted for different instants of time in Figs. 2.2 and 2.3. The mainly compressive forcing produces converging gas streams in random directions. At the interfaces in between these streams, the gas is strongly compressed and shock fronts are forming. As one can see in Fig. 2.2, the shock-compressed gas is very well tracked by AMR, while smooth regions remain coarse. Figure 2.3 shows the onset of turbulence due to instabilities at the collision interfaces. These structures are also covered by refined regions.

A quantitative indicator for the reliability of AMR are probability density functions (PDFs; see Sect. 1.2.2). Figures 2.4, 2.5 and 2.6 compare PDFs of the mass density, the vorticity modulus, and divergence after one dynamical time scale for AMR and a static-grid simulation with uniform grid resolution $N_0 = 768^3$. The agreement between the PDFs from these two simulations is very good. In particular, the high-density and high-vorticity tails are well reproduced with refinement by vorticity (Eq. 2.13) and compression rate (Eq. 2.14). The deviations in the left tails of the PDF are expected because the refinement criteria are optimized to capture peak values, which are located in the right tails. But important effects such as high overdensities that might give rise to gravitational collapse are associated with the

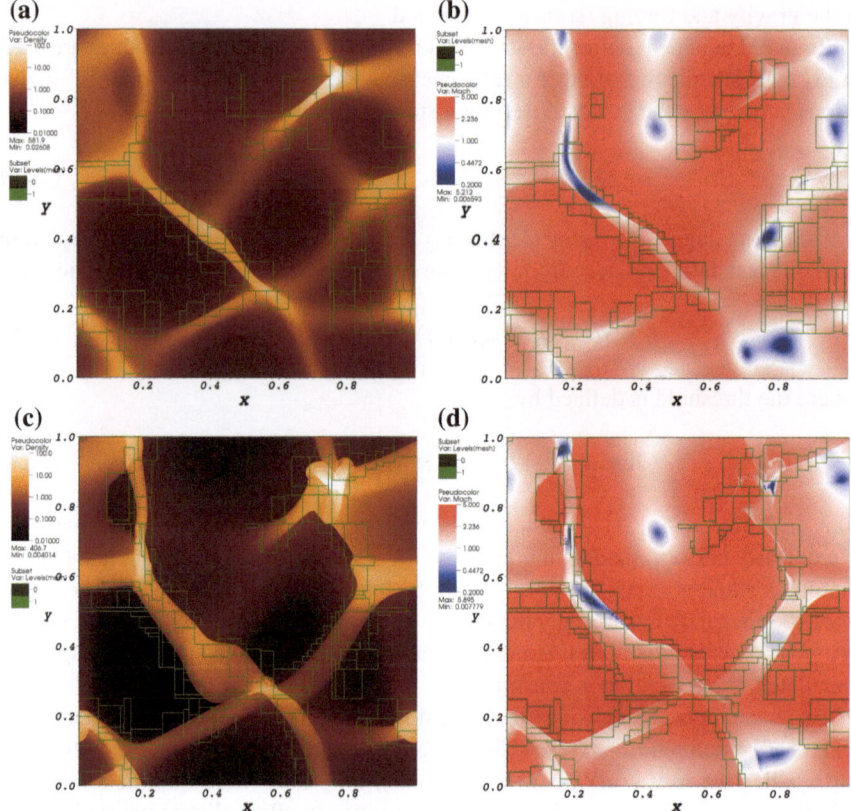

Fig. 2.2 Slices of the mass density ρ and the Mach number $\mathcal{M} = v/c_s$ in snapshots from an AMR simulation of supersonic turbulence [8]. The rectangles show the boundaries of refined grid patches **a** $\rho(t = 0.44T)$. **b** $\mathcal{M}(t = 0.44T)$. **c** $\rho(t = 0.58T)$. **d** $\mathcal{M}(t = 0.58T)$

right tails. The comparison demonstrates that, in principle, AMR can be applied to turbulence simulations, although the continuously changing grid structure is difficult to handle with sufficient efficiency. If turbulence is produced only in certain regions, however, AMR simulations can benefit from turbulence-based refinement criteria [32, 33].

2.3 Large Eddy Simulations

A common problem of astrophysical turbulence simulations are the usually very high Reynolds numbers. According to the criterion (1.24), direct numerical simulations, which encompass the full the dynamical range from the forcing length scale L to the viscous dissipation scale l_K, are infeasible. The majority of simulations are so-called

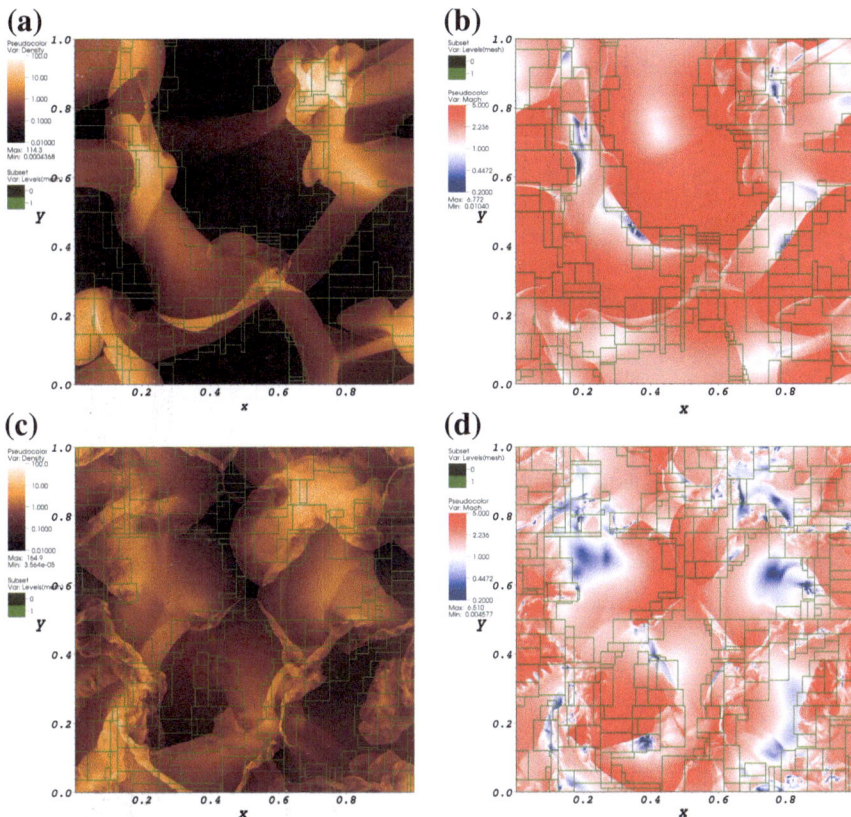

Fig. 2.3 Density and Mach number slices, continuing Fig. 2.2. **a** $\rho(t = 0.73T)$. **b** $\mathcal{M}(t = 0.73T)$. **c** $\rho(t = 1.02T)$. **d** $\mathcal{M}(t = 1.02T)$

implicit large eddy simulation (ILES). This means that the numerical viscosity associated with the truncation terms of the numerical discretization smoothes the flow over several grid cells and dissipates kinetic energy. Consequently, the physical scale of energy dissipation l_K is replaced by the grid scale Δ and the number of degrees of freedom is reduced to the number of grid cells. For example, all simulations mentioned in Sect. 2.1 belong to this category. On the other hand, large eddy simulations (LES) in the proper sense predict the rate at which energy is transported by the turbulent cascade from numerically resolved ($l \gtrsim \Delta$) to unresolved ($l \gtrsim \Delta$) length scales [34]. LES of compressible turbulence often use a subgrid scale model for the kinetic energy of numerically unresolved velocity fluctuations and the physical dissipation of energy (Eq. 1.13), which occurs on sub-resolution scales. The basic assumption of ILES is that the latter is approximated by the numerical dissipation.

To distinguish the exact solution of the compressible Navier-Stokes from the variables that are computed in LES, we use in the following the symbols $\overset{\infty}{\rho}$ for

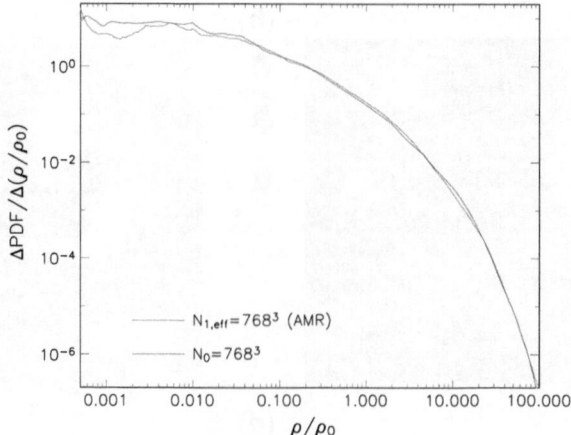

Fig. 2.4 Comparison of the probability density functions of the mass density ρ for static-grid and AMR simulations of mainly compressively driven turbulence with the same effective resolution (here, the acronym PDF in the vertical axis label signifies the probability distribution function, whose derivative is the probability density function) [8]

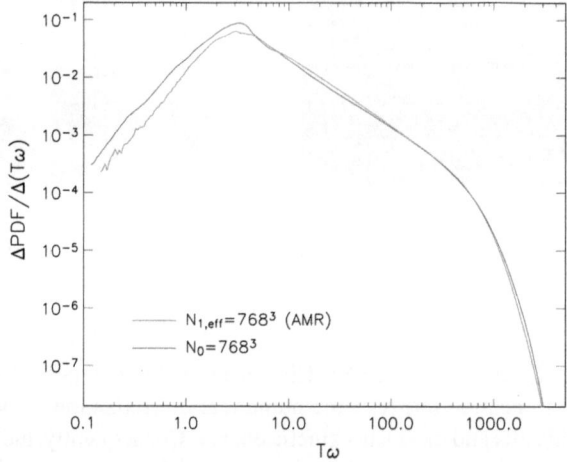

Fig. 2.5 Comparison of the probability density functions of the vorticity ω, as in Fig. 2.4

the mass density of the gas at infinite resolution, $\overset{\infty}{\mathbf{v}}$ for the exact flow velocity, etc. A consistent formulation of the dynamics on length scales $\gtrsim \Delta$ can by derived from the Navier-Stokes equations by means of the filter formalism introduced in [35]. The generalization of this formalism to compressible fluid dynamics is straightforward [36, 37]. The basic idea is to identify the numerically computed solution with the filtered variables $\rho := \langle \overset{\infty}{\rho} \rangle_\Delta$, $\mathbf{v} := \langle \overset{\infty}{\rho} \overset{\infty}{\mathbf{v}} \rangle_\Delta / \rho$, etc., where the filter operator $\langle \cdot \rangle_\Delta$ smoothes the variables $\overset{\infty}{\rho}$, $\overset{\infty}{\mathbf{v}}$, etc. over the length scale Δ. In LES, the filtering

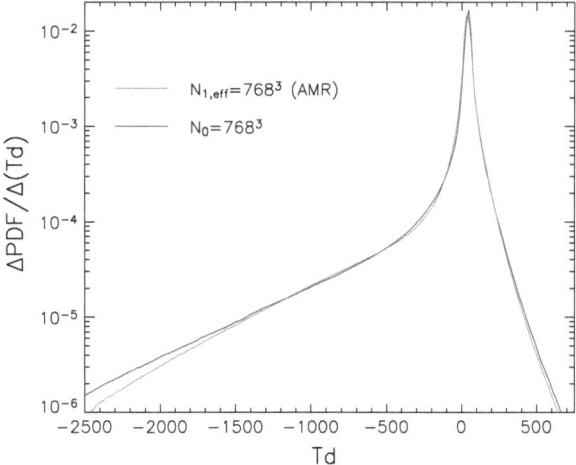

Fig. 2.6 Comparison of the probability density functions of the divergence d, as in Fig. 2.4

usually corresponds to the numerical discretization. The dynamical equations for the computable, filtered quantities are similar to the unfiltered equations, with additional terms that are related to the subgrid-scale dynamics on length scales $l < \Delta$.

Let us consider Eq. (1.2) for the momentum density of the fluid. For the exact solution, this equation reads

$$\frac{\partial}{\partial t} \left(\overset{\infty\infty}{\rho\, \mathbf{v}} \right) + \nabla \cdot \left(\overset{\infty\infty}{\rho\, \mathbf{v}} \otimes \overset{\infty}{\mathbf{v}} \right) = \overset{\infty}{\rho} \left(\overset{\infty}{\mathbf{g}} + \overset{\infty}{\mathbf{f}} \right) - \nabla \overset{\infty}{P} + \nabla \cdot \overset{\infty}{\sigma}, \qquad (2.18)$$

where the viscous dissipation tensor $\overset{\infty}{\sigma}$ is given by

$$\overset{\infty}{\sigma}_{ij} = 2\nu \overset{\infty}{\rho} \left(\overset{\infty}{S}_{ij} - \frac{1}{3} \overset{\infty}{d}\, \delta_{ij} \right). \qquad (2.19)$$

Without loss of generality, we neglect the second viscosity.

By applying a homogeneous filter operator that is uniform in time, Eq. (2.18) is converted into an equation for the filtered momentum, $\rho\mathbf{v} = \langle \overset{\infty\infty}{\rho\, \mathbf{v}} \rangle$. This equation has the same form as the original equation, except for one term. Because of the non-linear advection term, the filtering introduces a stress term that accounts for the interaction between the numerically resolved flow and velocity fluctuations on subgrid scales:

$$\frac{\partial}{\partial t}(\rho\mathbf{v}) + \nabla \cdot (\rho\mathbf{v} \otimes \mathbf{v}) = \rho(\mathbf{g}+\mathbf{f}) - \nabla P + \nabla \cdot \left(\sigma + \tau_{\text{sgs}} \right), \qquad (2.20)$$

where the SGS turbulence stress tensor is defined by

$$\tau_{\text{sgs}} = -\langle \overset{\infty\infty}{\rho}\,\overset{\infty}{\mathbf{v}} \otimes \overset{\infty}{\mathbf{v}}\rangle_\Delta + \rho\mathbf{v} \otimes \mathbf{v}. \tag{2.21}$$

The second-order moment $\langle \overset{\infty\infty}{\rho}\,\overset{\infty}{\mathbf{v}} \otimes \overset{\infty}{\mathbf{v}}\rangle_\Delta$ is not explicitly computable in LES because the variations of the mass density $\overset{\infty}{\rho}$ and the velocity $\overset{\infty}{\mathbf{v}}$ below the grid scale are unknown. For this reason, an approximation in terms of filtered quantities has to be devised. This is the closure problem. Alternatively, $\langle \overset{\infty\infty}{\rho}\,\overset{\infty}{\mathbf{v}} \otimes \overset{\infty}{\mathbf{v}}\rangle_\Delta$ can be expressed in terms of higher-order moments, but this merely shifts the closure problem to the higher-order moments.

The SGS turbulence energy density is defined by the difference between the resolved kinetic energy and the filtered kinetic energy:

$$K_{\text{sgs}} := \frac{1}{2}\langle \overset{\infty\infty}{\rho}\,\overset{\infty}{\mathbf{v}} \cdot \overset{\infty}{\mathbf{v}}\rangle_\Delta - \frac{1}{2}\rho|\mathbf{v}|^2 = -\frac{1}{2}\text{tr}\,\tau_{\text{sgs}}, \tag{2.22}$$

where $\text{tr}\,\tau_{\text{sgs}} = \tau_{ii}$ is the trace of the SGS turbulence stress tensor (the components of τ_{sgs} are simply denoted by τ_{ij}). On the right-hand side of Eq. (2.20), the contribution from the trace corresponds to the term $-\frac{2}{3}\nabla K_{\text{sgs}}$. This term can be absorbed into the pressure gradient if the thermal pressure P is replaced by the effective pressure

$$P_{\text{eff}} = P + \frac{2}{3}K_{\text{sgs}} = P - \frac{1}{3}\text{tr}\,\tau_{\text{sgs}}. \tag{2.23}$$

The relative contribution of the turbulent pressure $P_{\text{sgs}} = \frac{2}{3}K_{\text{sgs}}$ compared to the thermal pressure P is characterized by the SGS turbulence Mach number

$$\mathcal{M}_{\text{sgs}} = \left(\frac{2K_{\text{sgs}}}{\rho c_{\text{s}}^2}\right)^{1/2}, \tag{2.24}$$

where c_{s} is the thermal speed of sound. $\mathcal{M}_{\text{sgs}}$ depends on the temperature of the fluid and the cutoff scale Δ. The dependence on Δ is investigated in Sect. 3.4.

The SGS turbulence energy is an intermediate reservoir of energy that exchanges energy with the resolved flow and loses energy through dissipation into heat. For the computation of K_{sgs}, the following partial differential equation has to be solved in addition to the filtered equations for the numerically resolved gas dynamics:

$$\frac{\partial}{\partial t}K_{\text{sgs}} + \nabla \cdot (\mathbf{v}K_{\text{sgs}}) = \Gamma + \Sigma - \rho(\varepsilon + \lambda) + \mathfrak{D}. \tag{2.25}$$

Here, $\Sigma = \tau_{ij}S_{ij}$ is the rate of SGS turbulence energy production by the turbulent cascade (also called the *turbulence energy flux*) and $\rho\varepsilon$ is the viscous dissipation rate smoothed over the grid scale. Effects caused by SGS fluctuations of the gravitational potential and the thermal pressure are given by Γ and $\rho\lambda$, respectively. The term $\mathfrak{D}$ accounts for SGS transport effects. Exact definitions of these terms are given in [36]. For our purpose it is sufficient to discuss closures of these terms, which are approximations in terms of the numerically resolved variables and K_{sgs}.

To compute the SGS turbulence stress tensor (2.21) in the highly compressible regime, the following closure is proposed in [37]:

$$\tau_{ij} = 2C_1\Delta(2\rho K_{sgs})^{1/2}S_{ij}^* - 2C_2 K_{sgs}\frac{2v_{i,k}v_{j,k}}{|\nabla \otimes \mathbf{v}|^2} - \frac{2}{3}(1 - C_2)K_{sgs}\delta_{ij}. \quad (2.26)$$

where $|\nabla \otimes \mathbf{v}| := (2v_{i,k}v_{i,k})^{1/2}$ is the norm of the resolved velocity derivative, $S_{ij}^* = S_{ij} - \frac{1}{3}d\delta_{ij}$ the trace-free part of S_{ij}, and d the divergence. The first term in Eq. (2.26) corresponds to the *eddy-viscosity closure*, which is commonly used for LES of weakly compressible turbulence [38]. The second term, which is non-linear in the velocity gradient, was originally applied to decaying adiabatic turbulence [39]. The standard eddy-viscosity closure with the turbulent viscosity

$$\nu_{sgs} = C_1\Delta(2K_{sgs}/\rho)^{1/2} \quad (2.27)$$

follows for $C_2 = 0$. In general, the linear eddy-viscosity term dominates if $(K_{sgs}/\rho)^{1/2}$ is small compared to $\Delta|S^*| \lesssim \Delta|\nabla \otimes \mathbf{v}|$. For strong turbulence intensity, i. e., $(K_{sgs}/\rho)^{1/2} \gtrsim \Delta|\nabla \otimes \mathbf{v}|$, the non-linear term contributes significantly. This particularly applies to intermittent events in supersonic turbulence, for which $\Delta|\nabla \otimes \mathbf{v}| \gtrsim c_s$. Independent of the values of C_1 and C_2, $\tau_{ii} = -2K_{sgs}$, as required by the identity (2.22). The trace-free part of the SGS turbulence stress tensor is denoted by τ_{ij}^*. In Sect. 3.4.1, the eddy-viscosity and non-linear contributions to the closure (2.26) are analyzed for supersonic turbulence.

The viscous stresses $\overset{\infty}{\sigma}_{ij}$ dissipate kinetic energy on the smallest dynamical length scales $l \sim l_K$ of the physical flow $\overset{\infty}{\mathbf{v}}$. The length scale l_K is called the Kolmogorov scale. The corresponding rate of energy dissipation, filtered over the grid scale Δ, is given by

$$\rho\varepsilon = \langle\overset{\infty}{\sigma}_{ij}\overset{\infty}{v}_{i,j}\rangle_\Delta = \langle 2\nu\overset{\infty}{\rho}\,\overset{\infty}{S}_{ij}^*\overset{\infty}{S}_{ij}^*\rangle_\Delta = \langle\nu\overset{\infty}{\rho}|\overset{\infty}{S}_{ij}^*|^2\rangle_\Delta. \quad (2.28)$$

It is important to note that $\rho\varepsilon \neq \sigma_{ij}v_{i,j}$, where σ_{ij} and $v_{i,j}$ are the filtered viscous stress tensor and the filtered velocity gradient, respectively. For fully developed incompressible turbulence, simple scaling arguments based on Eq. (1.24) show that the viscous stress term in the filtered momentum equation (2.20) is negligible if $l_K \ll \Delta$, i. e., $|\sigma| \ll |\tau_{sgs}|$ [40]. Consequently, the *physical* energy dissipation occurs entirely on subgrid scales $l \sim l_K \ll \Delta$. However, this does not imply that the rate of energy dissipation $\rho\varepsilon$ defined by Eq. (2.28) vanishes. In the limit $l_K/\Delta \to 0$, velocity fluctuations on ever smaller length scales give rise to arbitrarily steep velocity gradients, which add up to a non-vanishing product of the viscosity times the squared rate of strain, regardless of how small the viscosity is. This results in a non-zero, asymptotically constant mean rate of energy dissipation, which is supported by experimental and numerical evidences [41, 42]. LES of supersonic turbulence suggest that this assumption holds even for supersonic turbulence (see Sect. 3.4).

If we assume that SGS turbulence energy is dissipated into heat at a rate proportional to K_{sgs} divided by the time scale $\Delta(K_{\text{sgs}}/\rho)^{-1/2}$, the following closure for the dissipation rate is obtained:

$$\rho\varepsilon = C_\varepsilon \frac{K_{\text{sgs}}^{3/2}}{\rho^{1/2}\Delta}. \tag{2.29}$$

In the case of supersonic turbulence, however, no satisfactory closes are available for the pressure-dilatation $\rho\lambda$. The simplest solution is to neglect pressure dilatation [39]. The transport term in Eq. (2.25) can be modelled by the gradient-diffusion approximation [38]

$$\mathfrak{D} = \nabla \cdot \left[\kappa_{\text{sgs}} \nabla \left(\frac{K_{\text{sgs}}}{\rho} \right) \right], \tag{2.30}$$

where the SGS turbulent diffusivity is approximated by $\kappa_{\text{sgs}} \approx 0.65\Delta(\rho K_{\text{sgs}})^{1/2}$, as shown in [36].

Furthermore, we assume that the self-gravity of the gas has no significant effects on length scales below the grid resolution. Then $\Gamma = 0$ in Eq. (2.25). This corresponds to the condition that the local Jeans length $\lambda_{\text{J}} = c_{\text{s}}(\pi/G\rho)^{1/2}$ (see Sect. 1.4) is sufficiently large compared to the grid scale Δ, a condition that is usually satisfied in astrophysical simulations [43–45]. Therefore, the filtered equations of fluid dynamics resulting from the compressible Navier-Stokes equations in the limit of $l_{\text{K}} \ll \Delta$ read [36]

$$\frac{\partial}{\partial t}\rho + \nabla \cdot (\mathbf{v}\rho) = 0, \tag{2.31}$$

$$\frac{\partial}{\partial t}(\rho\mathbf{v}) + \nabla \cdot (\rho\mathbf{v} \otimes \mathbf{v}) = \rho(\mathbf{g}+\mathbf{f}) - \nabla(P + P_{\text{sgs}}) + \nabla \cdot \tau_{\text{sgs}}^*, \tag{2.32}$$

$$\frac{\partial}{\partial t}E_{\text{tot}} + \nabla \cdot (\mathbf{v}E_{\text{tot}}) = \rho\mathbf{v} \cdot (\mathbf{g}+\mathbf{f}) - \nabla \cdot [\mathbf{v}(P + P_{\text{sgs}})] \tag{2.33}$$

$$+ \nabla \cdot (\mathbf{v} \cdot \tau_{\text{sgs}}^*) - \Sigma + \rho\varepsilon.$$

The numerically resolved energy density is $E_{\text{tot}} = \rho e_{\text{tot}}$, where e_{tot} is defined by Eq. (1.5). Since the resolved velocity $\mathbf{v}$ is unaffected by the physical viscosity on length scales $l \geq \Delta$, the above set of equations defines the compressible *Euler* equations for computational fluid dynamics in a physically meaningful and consistent way. These equations are supplemented by the equation of state (1.6), the SGS turbulence energy equation (2.25), and the Poisson equation for the gravitational potential (1.9). The pure compressible Euler equations without SGS terms, on the other hand, do not follow from the compressible Navier-Stokes equation in the limit of infinite Reynolds number. In this case, there is no viscous dissipation at all and, by definition, ε vanishes identically. This is a mathematical idealization that does not describe turbulent flows in nature.

ILES follow as a special case from the above approach if two basic assumptions are made. Firstly, the discretization of the compressible Euler equations introduces a dissipative leading error term $\mathfrak{D}_{num}$ in the momentum equation (2.31). Implicitly, this term is assumed to be equivalent to the SGS turbulence stress term $\nabla \cdot \tau_{sgs}$. The second assumption is that $\mathbf{v} \cdot \mathfrak{D}_{num} = -\rho\varepsilon$, i.e., kinetic energy on the resolved scales is directly dissipated into heat at a rate that approximates the viscous dissipation on unresolved length scales. This is referred to as numerical viscosity or numerical dissipation. Effectively, the following equations are solved in ILES:

$$\frac{\partial}{\partial t}\rho + \nabla \cdot (\mathbf{v}\rho) = 0, \tag{2.34}$$

$$\frac{\partial}{\partial t}(\rho\mathbf{v}) + \nabla \cdot (\rho\mathbf{v} \otimes \mathbf{v}) = \rho(\mathbf{g} + \mathbf{f}) - \nabla P + \mathfrak{D}_{num}, \tag{2.35}$$

$$\frac{\partial}{\partial t}E_{tot} + \nabla \cdot (\mathbf{v}E_{tot}) = \rho\mathbf{v} \cdot (\mathbf{g} + \mathbf{f}) - \nabla \cdot (\mathbf{v}P). \tag{2.36}$$

Despite the lack of a strict mathematical justification, ILES serves as a very useful approximation in numerous astrophysical applications. Many properties of compressible turbulence are investigated with ILES. Moreover, numerical dissipation in addition to the explicit dissipation defined by Eq. (2.29) cannot be avoided in LES if numerical schemes for compressible fluid dynamics such as the piecewise parabolic method (PPM) are applied.

As an example, Fig. 2.7 shows a visualization of K_{sgs} from an LES of supersonic turbulence with a resolution of 512^3 grid cells [37]. The parameters of this simulation were chosen to match the ILES with solenoidal forcing presented in Sect. 2.1. An LES with purely compressive forcing is shown Fig. 2.8. In both cases, the forcing

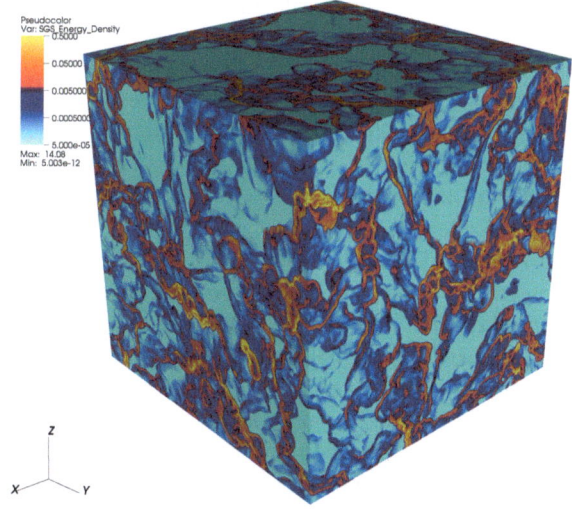

Fig. 2.7 Visualization of the SGS turbulence energy density K_{sgs} in a 512^3 LES with solenoidal forcing [37]

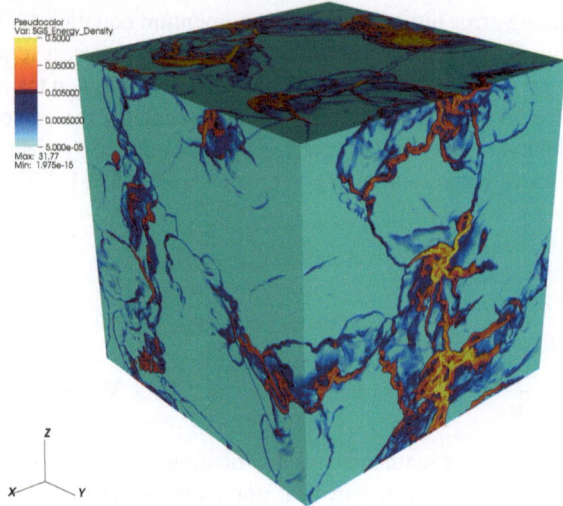

Fig. 2.8 Visualization of the SGS turbulence energy density K_{sgs} in a 512^3 LES with compressive forcing

magnitude was adjusted to obtain a steady-state RMS Mach number around 5. In the reddish regions of the plots, K_{sgs} is higher than the spatial mean, while it is lower in the bluish regions. The production of SGS turbulence energy depends on the numerically resolved turbulent velocity fluctuations. It is therefore interesting to compare K_{sgs} to the denstrophy $\Omega_{1/2} = \frac{1}{2}\left|\nabla \times \left(\rho^{1/2}\mathbf{u}\right)\right|^2$, which was proposed as an indicator of compressible turbulent velocity fluctuations [4]. The correlation diagrams of K_{sgs} versus $\Delta^2\Omega_{1/2}$ in Fig. 2.9 show that, on the average, $K_{\mathrm{sgs}} \sim 0.1\Delta^2\Omega_{1/2}$ for relatively large denstrophy values. Locally, however, K_{sgs} and $\Delta^2\Omega_{1/2}$ can substantially deviate from the average relation. This is a consequence of the different processes contributing to the SGS dynamics, which are not fully encompassed by the derivative of the resolved velocity field. For this reason, derived quantities such as the rate of strain or the denstrophy are only of limited utility to estimate effects of turbulence on unresolved length scales.

The effective pressure (2.23) is plotted versus the mass density in Fig. 2.10. One can see that the average of the effective pressure for a given mass density closely follows the isothermal relation $P \propto \rho$. This is because the mean turbulent pressure P_{sgs} is small compared to the thermal pressure for the resolution $\Delta = L/256$. However, the intermittency of turbulence can locally produce an effective pressure that exceeds the thermal pressure by one order of magnitude. For lower numerical resolution, this effect becomes stronger. In addition to the turbulent pressure, the diffusive non-diagonal viscous stresses τ_{ij}^* act on the resolved flow (see Eq. 2.32).

The rate of energy dissipation on the grid scale is predicted by Eq. (2.29). In ILES, one could alternatively extrapolate the expression for the microscopic dissipation rate (Eq. 1.13) to the grid scale Δ. This leads to the expression

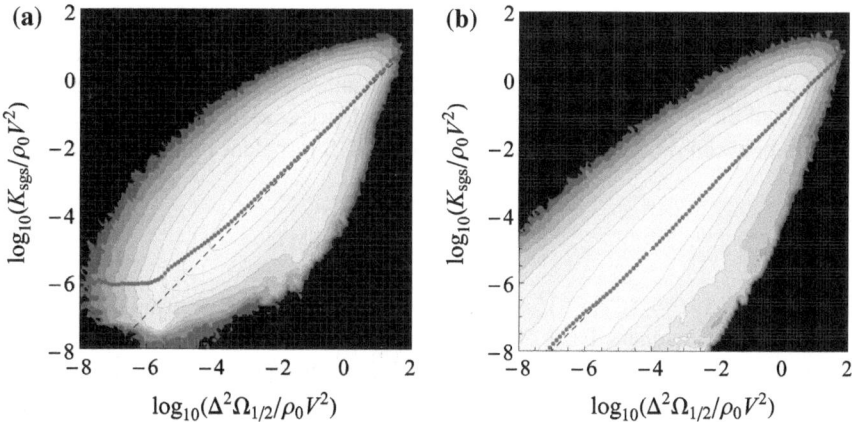

Fig. 2.9 Correlation diagrams of the SGS turbulence energy versus the denstrophy for 512^3 LES with solenoidal and compressive forcing [37]. The quantities are normalized by characteristic scales. The spacing of the contours is logarithmic. The average relation between both quantities is indicated by the *thick dotted lines*, and the *dashed line* shows the relation $K_{sgs} \sim 0.1 \Delta^2 \Omega_{1/2}$. **a** Solenoidal. **b** Compressive

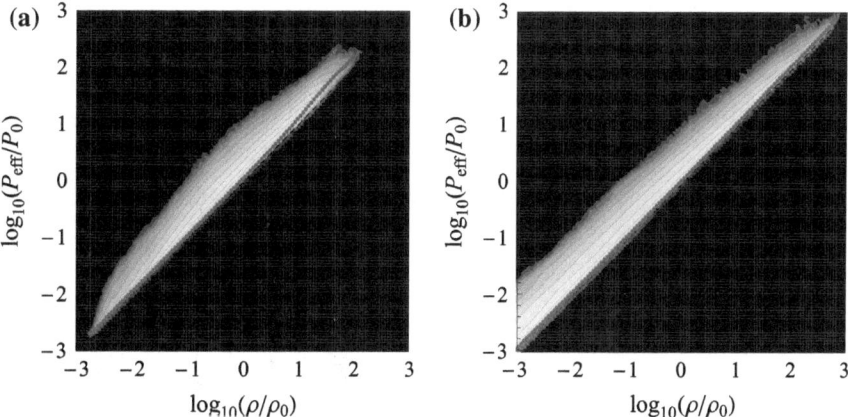

Fig. 2.10 Phase diagrams of the effective pressure defined by Eq. (2.23) versus the mass density for 512^3 LES with solenoidal and compressive forcing [37]. The averages of the SGS turbulence energy for particular values of the denstrophy are indicated by the *thick dotted lines*. **a** Solenoidal. **b** Compressive

$$\rho \varepsilon = \rho \nu_\Delta |S^*|^2, \qquad (2.37)$$

where a constant numerical viscosity $\nu_\Delta = V L / \mathrm{Re}_\Delta$ is assumed [46]. The Reynolds number on the grid scale can be estimated from mean squared rate of strain and the root mean square velocity: $Re = 2L^2 \langle |S^*|^2 \rangle / v_{rms}^2$.[1] The problem with this approach is that the viscosity on the grid scale, which corresponds to the eddy-viscosity (2.27) of the SGS model, cannot be assumed to be constant. Neglecting diffusion, compressibility, and the non-linear term in the SGS turbulence stress tensor (2.26), the equilibrium between production and dissipation of SGS turbulence energy in Eq. (2.25) implies

$$K_{sgs} \sim \frac{C_1}{C_\varepsilon} \rho \Delta^2 |S^*|^2.$$

Hence, $\varepsilon \sim (\Delta/C_\varepsilon)^2 (C_1 |S^*|)^3$ according to Eq. (2.29), which determines the rate of energy dissipation on the basis of the scale-separation of fluid dynamics. Comparing to Eq. (2.37), we see that $v_\Delta \sim \Delta^2 |S^*|$, which is not constant. This is a consequence of the fact that $\rho\varepsilon \neq \sigma_{ij} u_{i,j} \propto |S^*|^2$ (see Eq. 2.28). The discrepancy becomes apparent in Fig. 2.11, which shows correlation diagrams of the rate of energy dissipation calculated via Eq. (2.37) versus ε following from the SGS model. For low values of ε, one can see an average relation close to $|S^*|^2 \propto \varepsilon^{2/3}$. This is just what follows from the above estimate of the equilibrium dissipation rate. This behavior is reasonable because the neglected contribution of the non-linear term in Eq. (2.26) is relatively small for low values of K_{sgs}.

In computational astrophysics, the LES approach outlined in this section was applied to turbulent combustion in thermonuclear supernovae and to cosmological structure formation. In the former case, an SGS model is required to determine the turbulent diffusivity, which becomes much greater than the thermal diffusivity. This

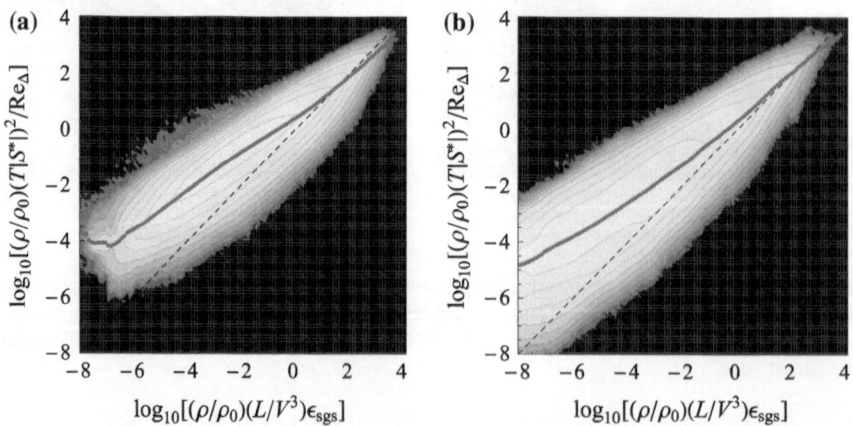

Fig. 2.11 Correlation diagrams of the normalized rate of energy dissipation defined by Eq. (2.37), where v_Δ assumes a constant value that is given by the numerical Reynolds number, versus the rate of energy dissipation (2.29) that is predicted by the SGS model [37]. The averages of expression (2.37) for given values of $\rho\varepsilon_{sgs}$ are indicated by the *thick dotted lines*. **a** Solenoidal. **b** Compressive

---

[1] See Eq. (3.4) in Sect. 3.1. For consistency with Eq. (2.37), however, ω_{rms}^2 is replaced by $\langle |S^*|^2 \rangle = \langle \omega^2 + \frac{4}{3} d^2 \rangle$.

leads to a significant enhancement of the nuclear burning rate [40, 48, 49]. To analyze the production of turbulence in simulations of cosmological structure formation, LES was combined with AMR [47, 50]. Moreover, an influence of the SGS model on the fragmentation properties of collapsing atomic cooling halos, from which the first stars in the Universe formed, is reported in [45].

References

1. R.S. Klessen, F. Heitsch, M.M. Mac Low, ApJ **535**, 887 (2000). doi:10.1086/308891
2. S. Boldyrev, A. Nordlund, P. Padoan, ApJ **573**, 678 (2002)
3. J. Ballesteros-Paredes, A. Gazol, J. Kim, R.S. Klessen, A.K. Jappsen, E. Tejero, ApJ **637**, 384 (2006). doi:10.1086/498228
4. A.G. Kritsuk, M.L. Norman, P. Padoan, R. Wagner, ApJ **665**, 416 (2007). doi:10.1086/519443
5. S. Dib, A. Brandenburg, J. Kim, M. Gopinathan, P. André, ApJ **678**, L105 (2008). doi:10.1086/588608
6. R. Kissmann, J. Kleimann, H. Fichtner, R. Grauer, MNRAS **391**, 1577 (2008). doi:10.1111/j.1365-2966.2008.13974.x
7. S.S.R. Offner, R.I. Klein, C.F. McKee, ApJ **686**, 1174 (2008). doi:10.1086/590238
8. W. Schmidt, C. Federrath, M. Hupp, S. Kern, J.C. Niemeyer, A&A **494**, 127 (2009). doi:10.1051/0004-6361:200809967
9. C. Federrath, J. Roman-Duval, R.S. Klessen, W. Schmidt, M. Mac Low, A&A **512**, A81+ (2010). doi:10.1051/0004-6361/200912437
10. D.C. Collins, A.G. Kritsuk, P. Padoan, H. Li, H. Xu, S.D. Ustyugov, M.L. Norman, ApJ **750**, 13 (2012). doi:10.1088/0004-637X/750/1/13
11. C. Federrath, R.S. Klessen, ApJ **761**, 156 (2012). doi:10.1088/0004-637X/761/2/156
12. E. Saury, M.A. Miville-Deschênes, P. Hennebelle, E. Audit, W. Schmidt, ArXiv e-prints (2013)
13. M.M. Mac Low, R.S. Klessen, A. Burkert, M.D. Smith, Phys. Rev. Lett. **80**, 2754 (1998)
14. J.M. Stone, E.C. Ostriker, C.F. Gammie, ApJ **508**, L99 (1998). doi:10.1086/311718
15. V. Eswaran, S.B. Pope, Comp. Fluids. **16**, 257 (1988)
16. W. Schmidt, W. Hillebrandt, J.C. Niemeyer, Comp. Fluids **35**, 353 (2006)
17. P.A. Davidson, *Turbulence: An Introduction for Scientists and Engineers* (Oxford University Press, Oxford, 2004)
18. A.G. Kritsuk, S.D. Ustyugov, M.L. Norman, P. Padoan, in Numerical modeling of space plasma flows, Astronum-2009, ed. by N.V. Pogorelov, E. Audit, G.P. Zank. *Astronomical Society of the Pacific Conference Series*, vol. 429. (San Francisco: Astronomical Society of the Pacific), p. 15 (2010)
19. R. Walder, D. Folini, A&AS **274**, 343 (2000). doi:10.1023/A:1026597318472
20. F. Heitsch, A.D. Slyz, J.E.G. Devriendt, L.W. Hartmann, A. Burkert, ApJ **648**, 1052 (2006). doi:10.1086/505931
21. E. Vázquez-Semadeni, G.C. Gómez, A.K. Jappsen, J. Ballesteros-Paredes, R.F. González, R.S. Klessen, ApJ **657**, 870 (2007). doi:10.1086/510771
22. P. Hennebelle, R. Banerjee, E. Vázquez-Semadeni, R.S. Klessen, E. Audit, A&A **486**, L43 (2008). doi:10.1051/0004-6361:200810165
23. M.J. Berger, J. Oliger, J. Chem. Phys. **53**, 484 (1984)
24. M.J. Berger, P. Colella, J. Chem. Phys. **82**, 64 (1989). doi:10.1016/0021-9991(89)90035-1
25. B.W. O'Shea, G. Bryan, J. Bordner, M.L. Norman, T. Abel, R. Harkness, A. Kritsuk, in Adaptive mesh refinement—theory and applications, ed. by T. Plewa, T. Linde, V.G. Weirs. *Lecture Notes in Computational Science and Engineering*, vol. 41 (Springer, Berlin, 2004), p. 341. http://esoads.eso.org/abs/2004astro.ph.3044O
26. B. Fryxell, K. Olson, P. Ricker, F.X. Timmes, M. Zingale, D.Q. Lamb, P. MacNeice, R. Rosner, J.W. Truran, H. Tufo, ApJS **131**, 273 (2000). doi:10.1086/317361

27. A.S. Almgren, J.B. Bell, M.J. Lijewski, Z. Lukić, E. Van Andel, ApJ **765**, 39 (2013). doi:10. 1088/0004-637X/765/1/39
28. R. Teyssier, A&A **385**, 337 (2002). doi:10.1051/0004-6361:20011817
29. S. Fromang, P. Hennebelle, R. Teyssier, A&A **457**, 371 (2006). doi:10.1051/0004-6361: 20065371
30. A.G. Kritsuk, M.L. Norman, P. Padoan, ApJ **638**, L25 (2006). doi:10.1086/500688
31. F. Vazza, G. Brunetti, A. Kritsuk, R. Wagner, C. Gheller, M. Norman, A&A **504**, 33 (2009). doi:10.1051/0004-6361/200912535
32. L. Iapichino, J. Adamek, W. Schmidt, J.C. Niemeyer, MNRAS **388**, 1079 (2008). doi:10.1111/ j.1365-2966.2008.13137.x
33. S. Paul, L. Iapichino, F. Miniati, J. Bagchi, K. Mannheim, ApJ **726**, 17 (2011). doi:10.1088/ 0004-637X/726/1/17
34. S.B. Pope, *Turbulent Flows* (Cambridge University Press, Cambridge, 2000)
35. M. Germano, J. Fluid Mech. **238**, 325 (1992)
36. W. Schmidt, J.C. Niemeyer, W. Hillebrandt, A&A **450**, 265 (2006). doi:10.1051/0004-6361: 20053617
37. W. Schmidt, C. Federrath, A&A **528**, A106+ (2011). doi:10.1051/0004-6361/201015630
38. P. Sagaut, *Large eddy simulation for incompressible flows: an introduction* (Springer, Berlin, 2006)
39. P.R. Woodward, D.H. Porter, S. Anderson, T. Fuchs, F. Herwig, J. Phys. Conf. Ser. **46**, 370 (2006). doi:10.1088/1742-6596/46/1/052
40. F.K. Röpke, W. Schmidt, in Interdisciplinary aspects of turbulence, ed. by W. Hillebrandt, F. Kupka, *Lecture Notes in Physics*, vol. 756. (Springer, Berlin, 2009), pp. 255-+
41. U. Frisch, *Turbulence. The legacy of A.N. Kolmogorov* (Cambridge University Press, Cambridge, 1995)
42. T. Ishihara, T. Gotoh, Y. Kaneda, Ann. Rev. Fluid Mech. **41**, 165 (2009). doi:10.1146/annurev. fluid.010908.165203
43. J.K. Truelove, R.I. Klein, C.F. McKee, J.H. Holliman II, L.H. Howell, J.A. Greenough, ApJ **489**, L179 (1997). doi:10.1086/316779
44. C. Federrath, S. Sur, D.R.G. Schleicher, R. Banerjee, R.S. Klessen, ApJ **731**, 62 (2011). doi:10. 1088/0004-637X/731/1/62
45. M.A. Latif, D.R.G. Schleicher, W. Schmidt, J. Niemeyer, MNRAS **430**, 588 (2013). doi:10. 1093/mnras/sts659
46. L. Pan, P. Padoan, A.G. Kritsuk, Phys. Rev. Lett. **102**(3), 034501 (2009). doi:10.1103/ PhysRevLett.102.034501
47. L. Iapichino, W. Schmidt, J.C. Niemeyer, J. Merklein, MNRAS **414**, 2297 (2011). doi:10.1111/ j.1365-2966.2011.18550.x
48. W. Schmidt, J.C. Niemeyer, W. Hillebrandt, F.K. Röpke, A&A **450**, 283 (2006). doi:10.1051/ 0004-6361:20053618
49. F.K. Röpke, W. Hillebrandt, W. Schmidt, J.C. Niemeyer, S.I. Blinnikov, P.A. Mazzali, ApJ **668**, 1132 (2007). doi:10.1086/521347
50. A. Maier, L. Iapichino, W. Schmidt, J.C. Niemeyer, ApJ **707**, 40 (2009). doi:10.1088/0004-637X/707/1/40

Chapter 3
Turbulent Velocity Statistics

The application of stochastic forcing makes it possible to analyze the velocity and density statistics of statistically stationary isotropic turbulence in the highly compressible regime. If the forcing is strong enough to produced supersonic flow velocities, the dissipation of the kinetic energy rapidly heats the gas, which entails a continuously rising speed of sound. For this reason, supersonic turbulence in a steady state with constant mean Mach number cannot be maintained without removing the excess heat. This can be realized either by explicit cooling or by artificially increasing the internal energy to very large values, while keeping the thermal pressure sufficiently low. As pointed out in Sect. 1.2.2, isothermal gas dynamics follows from adiabatic gas dynamics in the limit $\gamma \rightarrow 1$, corresponding to a diverging internal energy. This can be numerically implemented by setting γ to a value very close to unity. A different approach would be to directly solve the reduced set of Eqs. (1.30) and (1.31) for isothermal gas dynamics. For the discussion of turbulence statistics in this and the following chapter, three simulations of supersonic turbulence with different values of the Helmholtz decomposition parameter ζ in Eq. (2.9) play a central role. We will refer to these simulations as ENZO_HYBR, FLASH_SOLN, and FLASH_COMP. Turbulence is driven by purely solenoidal ($\zeta = 1.0$ in Eq. 2.9) and compressive ($\zeta = 0.0$) forcing in FLASH_SOLN and FLASH_COMP [1], respectively, while ENZO_HYBR uses mostly compressive forcing with $\zeta = 0.1$ [2]. As the names suggest, the first two simulations were performed with the FLASH code [3] and the latter with ENZO [4]. Apart from that, the typical Mach numbers differ. All three simulations have in common that they are implicit large eddy simulations (ILES; see Sect. 2.3). In this chapter, we focus on the scaling laws following from two-point velocity statistics and intermittency properties. In addition, the energy flux through the turbulent cascade is calculated. Thereby, subgrid scale (SGS) model coefficients Sect. 2.3 are calibrated. The resulting SGS model is applied in explicit large eddy simulations (LES) to investigate the dependence of the energy fraction below a given length scale (the grid resolution), which not only verifies the consistency of the model, but can be regarded as an alternative formulation of the second-order scaling of compressible turbulence.

W. Schmidt, *Numerical Modelling of Astrophysical Turbulence*, 41
SpringerBriefs in Astronomy, DOI: 10.1007/978-3-319-01475-3_3,
© The Author(s) 2014

3.1 Global Averages and Probability Density Functions

The production of turbulence driven by the mostly compressive force in the simulation ENZO_HYBR is illustrated by the time sequence of three-dimensional renderings of the vorticity modulus ω in Fig. 3.1.[1] Initially, high vorticity mainly occurs at shock fronts, which appear as bent sheets and bulge-like structures. These structures subsequently break up into smaller shocklets and, particularly at the vertices of shocks, turbulent eddies are developing. The extremely intricate turbulent flow resulting from

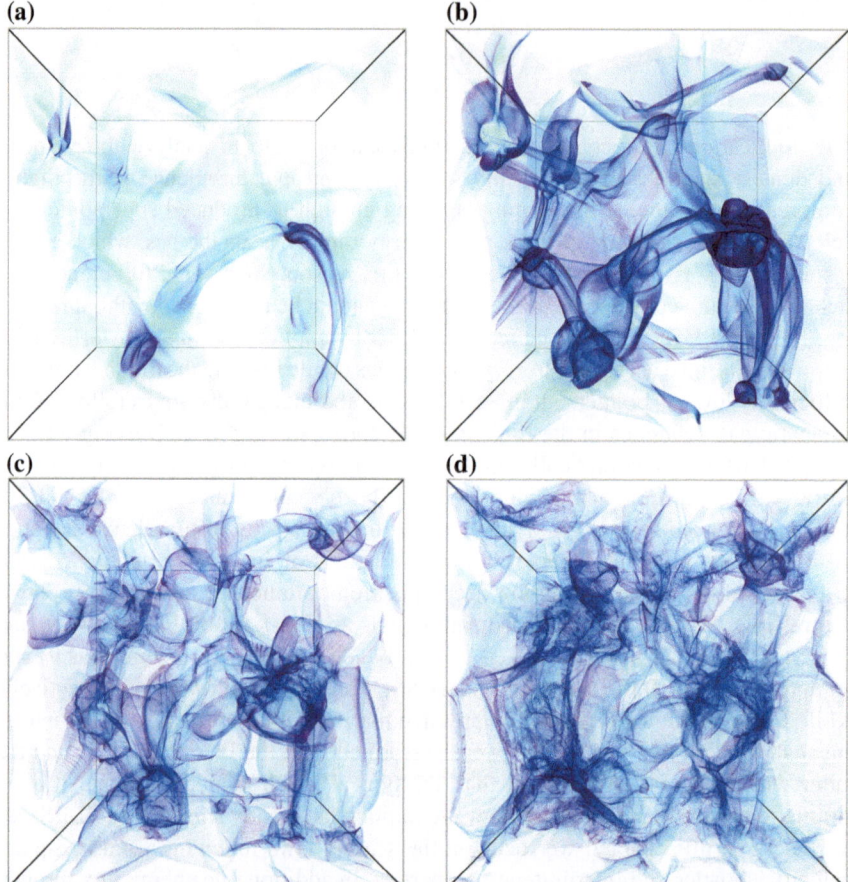

(a)   **(b)**

(c)   **(d)**

Fig. 3.1 Volume renderings of the vorticity modulus ω for compressively driven turbulence with $\zeta = 0.1$ [2]. **a** $t = 0.44T$ **b** $t = 0.58T$ **c** $t = 0.73T$ **d** $t = 0.87T$

---

[1] The volume renderings were made from the AMR run, while the statistics was computed from the uniform-grid run [2]. Although it is demonstrated in Sect. 2.2 that the statistics of the AMR run is closely matched by the uniform-grid run, only the latter was continued over several integral time scales.

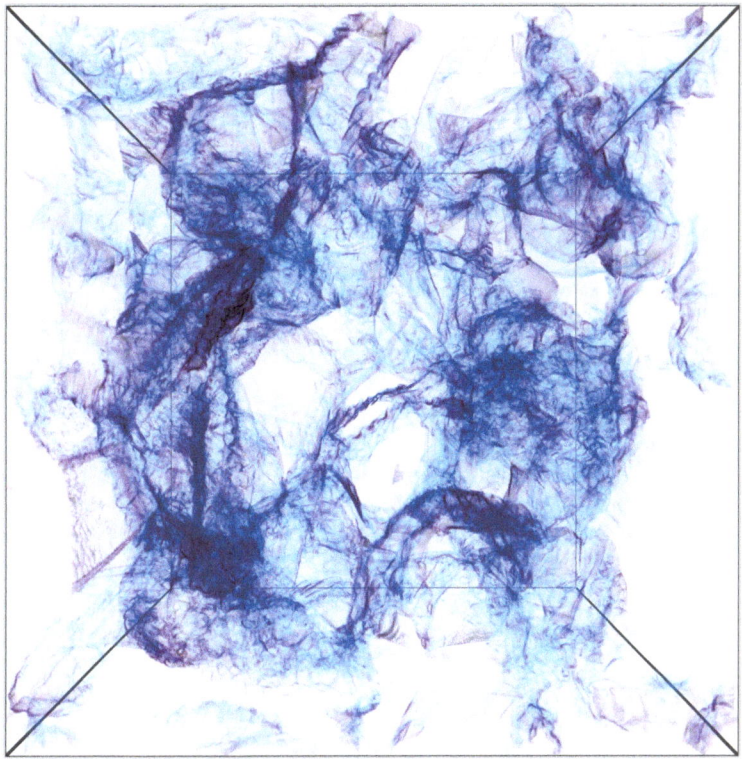

Fig. 3.2 Volume rendering of the vorticity modulus ω after one integral time ($t = 1.02T$), continuing the evolution shown in Fig. 3.1

this process can be seen in detail in Fig. 3.2. The vorticity is growing rapidly over the first integral time T, as can be seen from the RMS values $\omega_{\text{rms}} = \langle \omega^2 \rangle^{1/2}$ listed in Table 3.1. All quantities in this table are normalized such that the mean density $\rho_0 - 1$, the initial pressure $P_0 = 1$, and the size of the computational domain $X = 2L = 1$. The cumulative probabilities of Mach numbers greater than unity indicate that more than 90 % of the domain is filled by supersonic flow for all instants shown in Figs. 3.1 and 3.2. The main parameter of supersonic isothermal turbulence is the RMS Mach number

$$\mathcal{M}_{\text{rms}} = \left\langle (v/c_{\text{s}})^2 \right\rangle^{1/2}, \tag{3.1}$$

where $c_{\text{s}}^2 = \gamma P/\rho \simeq P/\rho$ is the squared speed of sound. Around $t \approx 0.5T$, $\mathcal{M}_{\text{rms}}$ reaches its largest values and also the highest densities occur due to strong shock compressions. From $t \approx 1.0T$ onwards, ω_{rms} saturates and $\mathcal{M}_{\text{rms}}$ gradually decreases because of the slow adiabatic heating for $\gamma = 1.01$. Since $\rho e_{\text{int}} = 100P$ and the mean energy dissipation rate is $\sim \rho e_{\text{kin}}/T \sim 3P/T$ (see Eq. 1.5), the relative change of the internal energy is $\sim (e_{\text{kin}}/e_{\text{int}})/T \approx 0.03/T$. Nevertheless, the flow can be

Table 3.1 Global statistics at different instants of time

t/T	ρ_{max}	$\sigma(\rho)$	$P(v/c_s > 1)$	$\mathcal{M}_{rms}$	ω_{rms}	λ/Δ	λ_{mw}/Δ	r_{cs}
0.15	2.3	0.21	0.44	1.04	2.5	707.1	687.6	0.97
0.29	76.9	1.05	0.86	1.92	5.4	656.4	382.5	0.97
0.44	1335.0	3.72	0.93	2.61	28.8	154.9	37.5	0.92
0.58	495.4	3.71	0.95	2.92	86.1	57.9	20.8	0.87
0.73	336.0	3.34	0.96	2.80	137.9	34.8	16.4	0.78
0.87	441.1	2.95	0.95	2.63	168.5	27.0	13.6	0.67
1.02	179.4	2.46	0.93	2.49	187.3	23.1	13.3	0.58
2.03	128.8	2.65	0.93	2.50	201.0	21.7	12.1	0.51
3.05	186.3	2.24	0.92	2.44	204.4	21.3	12.8	0.47
3.92	259.7	2.31	0.92	2.35	207.6	20.5	13.0	0.47
5.08	268.4	2.99	0.91	2.31	208.8	20.2	11.5	0.48
5.95	172.4	2.24	0.91	2.23	202.1	20.5	12.4	0.47
9.14	383.0	3.05	0.91	2.20	210.5	20.2	12.0	0.47

From left to right, the columns contain the normalized time, peak density, standard deviation of the density, cumulative probability of Mach numbers >1, RMS Mach number, RMS vorticity, mass-weighted RMS vorticity, Taylor scale, mass-weighted, Taylor scale, small-scale compressive ratio and the corresponding mass-weighted compressive ratio

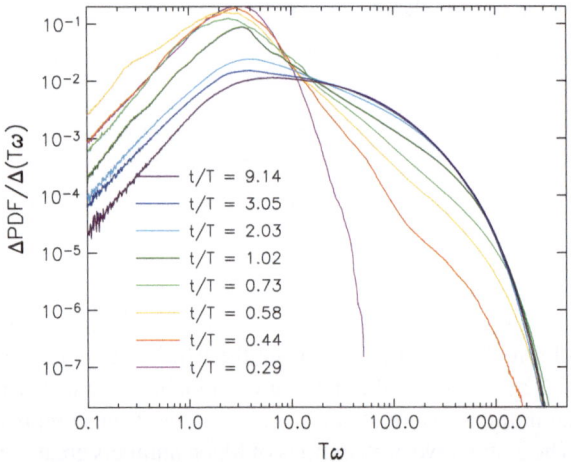

Fig. 3.3 Temporal evolution of the probability density functions of the vorticity modulus ω (here, the acronym PDF in the vertical axis label signifies the probability distribution function, whose derivative is the probability density function) [2]

regarded as quasi-isothermal, as $\mathcal{M}_{rms}$ changes significantly only over many integral time scales.

The rapid evolution from initially smooth to turbulent flow within the first integral time becomes manifest in the broadening of the probability density function (PDF) of the vorticity modulus plotted in Fig. 3.3 (see Sect. 1.2.2 for a general description of PDFs). At $t \approx 0.5T$, a power-law decrease of the ω-PDF in the subrange

$10 \lesssim \omega \lesssim 100$ becomes apparent, which might arise from a self-similar ensemble of shocks. The subsequent flattening of the power-law subrange and the development of an exponential tail indicates the transition from a transient shock-dominated phase to the statistically stationary turbulent phase, in which vortices play a prominent role. From $t \approx 2.0T$ onwards, the PDFs of ω show only little variation in time.

To investigate whether turbulence settles into a steady state, the Taylor scale λ is an important indicator. For incompressible turbulence, the expression

$$\lambda = \sqrt{5}\frac{v_{rms}}{\omega_{rms}}, \tag{3.2}$$

follows from the two-point autocorrelation of the turbulent velocity field [5]. Basically, the Taylor scale is the product of the time scale given by the inverse RMS vorticity, $1/\omega_{rms}$, and the RMS velocity fluctuation. If the above expression is tentatively applied to compressible turbulence, the values listed in Table 3.1 are obtained. The flow that is initially induced by the forcing is smooth and varies only over the integral length L. Thus, $\lambda \sim L$ at early time. Subsequently, the Taylor scale drops substantially, the reason being that the vorticity rises more rapidly than the velocity magnitude, as turbulent velocity fluctuations are developing on decreasing length scales. After about two integral time scales, the Taylor scale settles around $\lambda \approx 20$, which implies that the statistically stationary regime with a balance between energy injection by the forcing (v_{rms}) and small-scale dissipation (ω_{rms}) is reached. To account for compressibility, the kinetic energy density, $\rho e_{kin} = \frac{1}{2}\rho v^2$, and the enstrophy density, $\Omega = \frac{1}{2}\rho\omega^2$ can be averaged to obtain a mass-weighted Taylor scale:

$$\lambda_{mw} = \sqrt{\frac{5\langle\rho e_{kin}\rangle}{\langle\Omega\rangle}} \simeq \sqrt{\frac{5\langle\mathcal{M}^2 P\rangle}{\langle\rho\omega^2\rangle}}. \tag{3.3}$$

The values of λ_{mw} listed in Table 3.1 show a faster decline during the production phase in comparison to λ, with an asymptotic ratio $\lambda_{mw}/\lambda \approx 0.6$. This suggests that a significant fraction of the vorticity originates from shock-compressed gas.

The *effective* Reynolds number of an ILES can be estimated from the relation [6],

$$\frac{\lambda}{L} = \sqrt{\frac{10}{Re_{eff}}}, \tag{3.4}$$

without explicit knowledge of the numerical viscosity. By substituting $L = 384\Delta$ and $\lambda/\Delta \approx 20$, it follows that $Re_{eff} \approx 3.7 \cdot 10^3$. This value agrees well with the estimate based on Eq. (1.24), $Re_{eff} = (L/\Delta)^{4/3} \approx 2.8 \cdot 10^3$. A Reynolds number of a few 10^3, however, implies that the turbulent cascade is only barely resolved. This remains a severe limitation of three-dimensional simulations. Even the largest contemporary simulations with a resolution of 4096^3 [7] do not go beyond an effective Reynolds numbers of the order 10^4.

A statistical measure of the relative importance of compression effects is the so-called small-scale compressive ratio [8], which is also listed in Table 3.1:

$$r_{cs} = \frac{d_{rms}^2}{d_{rms}^2 + \omega_{rms}^2}. \tag{3.5}$$

At early time, this ratio is close to unity, i. e., the rotation-free, compressive component of the velocity dominates. This reflects the properties of the large-scale force. With growing vorticity, r_{cs} decreases. The compressive ratio in the steady state is slightly less than 0.5. For comparison, $r_{cs} \approx 0.28$ was found for a forcing field with about $1/3$ compressive and $2/3$ solenoidal modes [9].

3.2 Two-Point Statistics of the Velocity

For supersonic turbulence in molecular clouds, usually transversal structure functions are considered [2, 9]. Observationally, transversal velocity fluctuations follow from measurements of the line broadening at positions separated by a certain angular distance. Here, we define the structure functions in terms of the velocity increments $\delta \mathbf{v} = \mathbf{v}(\mathbf{x}, t) - \mathbf{v}(\mathbf{x} + \mathbf{l}, t)$ projected perpendicular to $\mathbf{l}$:

$$S_p^{\perp}(l) = \langle |\delta v^{\perp}(l)|^p \rangle, \quad \text{where} \quad \delta v_i^{\perp} = \left(\delta_{ij} - \frac{l_i l_j}{l^2} \right) \delta v_j. \tag{3.6}$$

Since the computation of statistically converged structure functions of higher order is a computationally extremely demanding task, a Monte-Carlo algorithm is applied to average velocity differences from randomly sampled points $\mathbf{x}$ and $\mathbf{x} + \mathbf{l}$ over bins of spatial separation l [1, 2].

3.2.1 Absolute Scaling Exponents

The second- and third-order structure functions plotted in Fig. 3.4 were computed from approximately 2.4×10^{10} data pairs of the simulation ENZO_HYBR [2]. Also shown in Fig. 3.4 are power law fits in the subrange $8 \leq l/\Delta \leq 50$ for $t/T \approx 1.0$ and $8 \leq l/\Delta \leq 70$ for $t/T \geq 2$. The resulting scaling exponents ζ_p for $p = 1, \ldots, 5$ are listed in Table 3.2 for different simulation snapshots. As one can see, $S_2^{\perp}$ is steeper at time $t/T \approx 1.0$ in comparison to later instants of time, but then turbulence reaches the statistically stationary regime, in agreement with the global averages and PDFs investigated in Sect. 3.1. However, the second-order scaling law with the exponent $\zeta_2 \approx 0.87$ remains much stiffer than the Kolmogorov two-third law and the exponent ζ_3 is greater than unity by more than two standard deviations for most data points. The inertial-range scaling of the RMS velocity fluctuation,

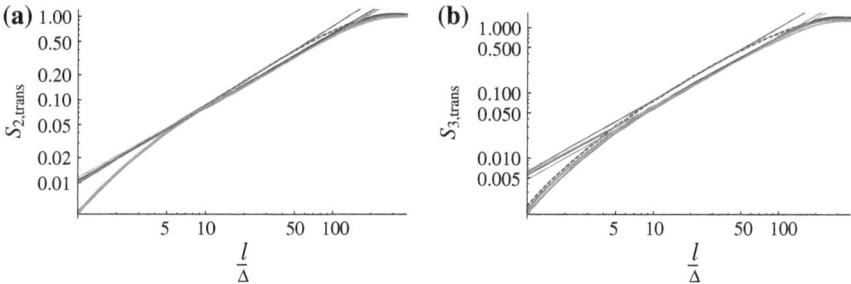

Fig. 3.4 Second-order **a** and third-order **b** transversal structure functions for $t/T \approx 1.0$ (*dashed lines*) and $t/T \approx 2.0, 3.0, 4.0$ (*solid lines*) [2]. Also shown are the corresponding power-law fit functions (*thin lines*). The structure functions are normalised by V^p, where V is the characteristic velocity of the forcing (see Eq. 2.11)

Table 3.2 Scaling exponents ζ_p obtained from power-law fits to the structure functions $S_p^{\perp}(l)$ for several stages of the turbulent flow evolution

t/T	$p = 1$	$p = 2$	$p = 3$	$p = 4$	$p = 5$
1.02	0.654 ± 0.002	0.950 ± 0.003	1.107 ± 0.006	1.202 ± 0.009	1.256 ± 0.012
2.03	0.562 ± 0.004	0.888 ± 0.003	1.066 ± 0.008	1.167 ± 0.013	1.228 ± 0.016
3.05	0.538 ± 0.005	0.884 ± 0.003	1.097 ± 0.006	1.239 ± 0.013	1.344 ± 0.019
4.06	0.531 ± 0.003	0.858 ± 0.004	1.035 ± 0.013	1.123 ± 0.023	1.163 ± 0.031
4.93	0.530 ± 0.002	0.848 ± 0.006	1.018 ± 0.016	1.100 ± 0.026	1.135 ± 0.034
5.95	0.523 ± 0.003	0.852 ± 0.004	1.044 ± 0.012	1.157 ± 0.021	1.224 ± 0.027
6.96	0.546 ± 0.004	0.875 ± 0.004	1.053 ± 0.010	1.150 ± 0.017	1.208 ± 0.021

$$\delta v_{\mathrm{rms}}(l) \simeq \sqrt{S_2^{\perp}(l)} \propto l^{\alpha},$$

is given by the exponent $\alpha = \zeta_2/2$. The average exponent ζ_2 for $t/T \geq 2$ corresponds to $\alpha \approx 0.43$. The scaling behavior of the squared velocity fluctuation can also be specified by the the slope of the turbulence energy spectrum function $E(k) \propto k^{-\beta}$ (see Eq. 1.20). In [2], the time-averaged value $\beta \approx 1.89$ is obtained from least-square fits to the energy spectra for $t/T \geq 2$. Theoretically, $\beta = 1 + \zeta_2$ for isotropic turbulence. Thus, the implied value of ζ_2 agrees well with the result from the calculation of the second-order structure function.

Observations of molecular clouds indicate power-law relations between the measured velocity dispersion and the size, $v_l \propto l^{\alpha}$. The steady-state value of α inferred from the simulation ENZO_HYBR is close to some observational values, for example, $\alpha \approx 0.4$ in [10] and 0.43 in [11]. For the Perseus cloud, a spectral index $\beta \approx 1.81$, corresponding to $\alpha \approx 0.4$, is reported in [12]. A principal-component analysis of measurements on various molecular clouds [13], on the other hand, yields a mean value $\alpha \approx 0.57$, which is significantly higher than the numerical value. However, the exponents for individual molecular clouds vary in the range $\alpha \approx 0.33 \ldots 0.81$, which encompasses the simulation result.

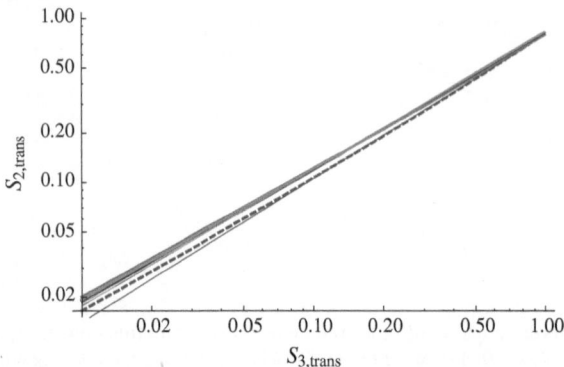

Fig. 3.5 Third-order versus second-order transversal structure functions for the same instants of time as in Fig. 3.4 [2]

For the FLASH simulations with solenoidal and compressive forcing, a detailed comparison to observations is made in [1]. This article addresses the problem that the full three-dimensional structure of turbulent velocity fields obtained from numerical simulations is not observable. This is basically a consequence of the unavoidable projection effects in astronomical observations. It is possible, however, to apply techniques such as the principal-component analysis to numerical data in order to mimic observational measurements. The results of this study suggest that the systematic differences in the scaling exponents, which are found for solenoidal and compressive forcing, can also be seen in observed objects. An interpretation of these results is that the observed scaling exponents for molecular-cloud turbulence result from a combination of different driving mechanisms, which varies from cloud to cloud.

3.2.2 Relative Scaling Exponents

In comparison to $S_2^\perp(l)$ and $S_3^\perp(l)$, a plot of $S_2^\perp$ versus $S_3^\perp$ for the simulation ENZO_HYBR (see Fig. 3.5) shows an extended range, in which the relation is close to a power law. This property is known as extended self-similarity (ESS) [14]. The time-averaged transversal structure functions $S_p^\perp$ for $p = 1, \ldots, 5$ are plotted against $S_3^\perp$ in Fig. 3.6. It appears that ESS is very well satisfied, albeit with a somewhat smaller range for $p = 5$. The slopes of the corresponding power-law fits for $0.04 \le S_3^\perp \le 1.0$ yield the relative scaling exponents $Z_p = \zeta_p/\zeta_3$. This can be seen by eliminating the dependence on l from the power laws $S_p^\perp(l) \propto l^{\zeta_p}$ and $S_3^\perp(l) \propto l^{\zeta_3}$ so that

$$\log S_p^\perp = Z_p \log S_2^\perp + \text{const.}$$

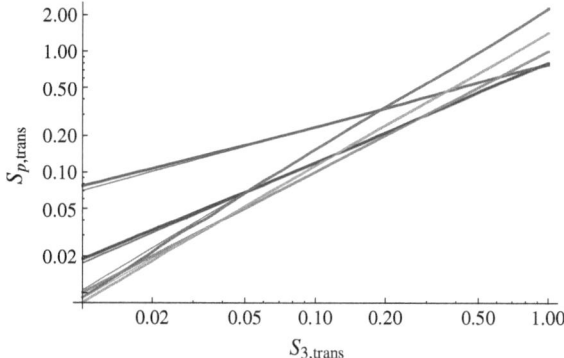

Fig. 3.6 ESS plot of transversal structure functions (*thick lines*) and the corresponding power-law fits (*thin lines*) up to fifth order at time $t \approx 2.03T$ [2]

Table 3.3 Comparison of the relative scaling exponents $Z_p = \zeta_p/\zeta_3$ predicted by the Kolmogorov theory and two intermittency models [15, 18] to the values obtained from various numerical simulations [2, 9, 16]

p	K41	$C = 2$, $\beta = 2/3$	$C = 1$, $\beta = 1/3$	ENZO [9]	FLASH_SOLN	ENZO_HYBR	FLASH_COMP
1	0.33	0.36	0.42	0.43	0.47	0.52	0.63
2	0.67	0.70	0.74	0.76	0.79	0.83	0.90
3	1.00	1.00	1.00	1.00	1.00	1.00	1.00
4	1.33	1.28	1.21	–	1.15	1.09	1.06
5	1.67	1.54	1.40	–	1.27	1.14	1.10

By averaging the values of Z_p for $t/T \geq 2$, the values summarized in Table 3.3 are obtained. The table also lists the corresponding values following from theoretical models and other simulations. In Fig. 3.7, the averaged scaling exponents are plotted with the corresponding scattering range for each p. While deviations from the incompressible Kolmogorov theory and the intermittency model defined by Eq. (1.51) [15] are expected, the values also differ significantly from the Kolmogorov-Burgers model (Eq. 1.53). The relative scaling exponents for FLASH_SOLN and FLASH_COMP are also inconsistent with this model [16]. Moreover, it can be seen that the values of Z_p for the forcing with $\zeta = 0.1$ (ENZO_HYBR) are in between the cases $\zeta = 0.0$ (FLASH_COMP) and $\zeta = 1.0$ (FLASH_SOLN). In contrast, good agreement with the Kolmogorov-Burgers model was found for a 1024^3 simulation performed with Enzo [9], although possible deviations of the scaling exponents from the Kolmogorov-Burgers model are discussed in [17].

There are several factors, from which systematic deviations in the scaling exponents might stem. Firstly, it is possible that the computation of the structure functions is not fully converged due to insufficient sampling. This might especially affect the scaling exponents of higher order. Increasing the sample size, however, did not

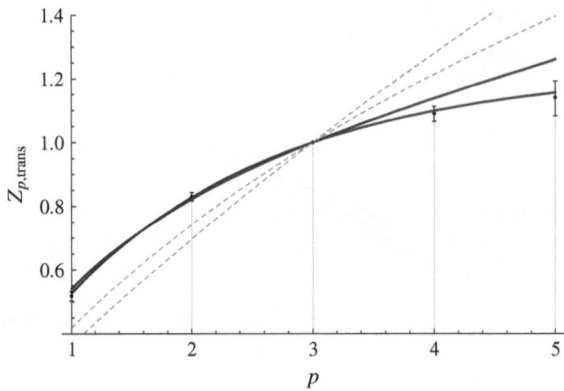

Fig. 3.7 Fits of the relative scaling exponents $Z_p = \zeta_p/\zeta_3$ to several models [2]. The error bars indicate the scatter of the numerically calculated exponents, the dots specify the averages. The dashed green and orange lines show the scalings predicted by the incompressible intermittency model (1.51) and the Kolmogorov-Burger model (1.53), respectively. Fits of the general intermittency model (1.52) yield the thick solid lines. The purple fit function, which closely matches the higher-order exponents, corresponds to $\Delta = 1$, whereas the blue function is obtained for $\Delta = 2/3$

change the obtained scaling exponents significantly. Secondly, the resolution of the simulation is possibly not high enough to allow for reliable estimates of the scaling exponents. However, other than the turbulence energy spectra, structure functions show a relatively broad power-law range, especially if ESS is applied. Runs with $\zeta = 0.0$ and $\zeta = 1.0$ at lower resolution show basically the same trends [16]. Thirdly, scaling exponents of transversal structure functions are considered, while theoretical model usually refer to longitudinal structure functions. But the numerical calculation of longitudinal structure functions shows even stronger disagreement of the higher-order scaling exponents with the Kolmogorov-Burgers model. Fourthly, the statistics might not be converged in time, but this is unlikely because neither the vorticity PDFs (see Fig. 3.3) nor the scaling exponents listed in Table 3.2 show systematic variations in time. Consequently, the scalings following from the simulations are probably sound.

An important questions regards the dependence on the degree of compressibility, as measured by the RMS Mach number. Given the relatively low RMS Mach number ($\mathcal{M}_{rms} \approx 2.2 \ldots 2.5$) of the simulation ENZO_HYBR compared to $\mathcal{M}_{rms} \approx 5$ for FLASH_SOLN and FLASH_COMP, it is likely that vortices contribute more in comparison to shocks in the former case. Based on results from MHD simulations with different Mach numbers, a one-parameter family of intermittency models for different Mach numbers is proposed in [19]. By assuming that $\Delta = 2/3$ is universal, the intermittency parameter $\beta = 1 - 2/(3C)$ is obtained from Eq. (1.50), where the co-dimension C decrease with $\mathcal{M}_{rms}$ from $C = 2$ in the weakly compressible regime to $C = 1$ for very high Mach numbers. The corresponding scaling exponents are

$$Z_p = \frac{p}{9} + C\left[1 - \left(1 - \frac{2}{3C}\right)^{p/3}\right]. \tag{3.7}$$

The She-Lévêque and Kolmogorov-Burgers models follow as limiting cases for $C = 2$ and $C = 1$, respectively. However, the best fit of this equation to the time-averaged scaling exponents Z_p in Fig. 3.7 yields $C \approx 0.71$, which falls outside the range of the one-parameter model. If the model were correct, the second-order exponent Z_2 should vary monotonically with the RMS Mach number from about 0.7 in the subsonic limit to the Kolmogorov-Burgers value 0.74 in the hypersonic limit. While $Z_2 \approx 0.76$ is found in [9], which is consistent with the Kolmogorv-Burgers exponent within the statistical uncertainty, $Z_2 \approx 0.83$ is obtained for ENZO_HYBR, although the RMS Mach number is smaller by at least a factor of two.

These discrepancies raise the question of whether the anomalous exponent $\Delta = 2/3$ for the most singular dissipative structures applies to the supersonic regime. This is the underlying assumption of Eq. (3.7) and its hypersonic limiting case, the Kolmogorov-Burgers model. By fitting the general two-parameter model (1.52) to the data, however, the preferred value of Δ is found to be around unity. How can this be understood? Let us consider Burgers turbulence [20], for which

$$v_l^2 \sim V^2 \frac{l}{L}.$$

The dissipation time scale is of the order of v_l^2 divided by the scale-invariant mean dissipation $\langle \varepsilon \rangle$, i. e., $\tau_l \sim (V^2/\langle \varepsilon \rangle)(l/L) \propto l$. By the same argument as in [15], the scaling of the most singular dissipative structures is then given by

$$\varepsilon_l^{(\infty)} \sim \frac{V^2}{\tau_l} \sim \langle \varepsilon \rangle \left(\frac{l}{L}\right)^{-1}.$$

Therefore, $\Delta = 1$ if these structures obey Burgers scaling, which is the scaling law of shock fronts. Therefore, the two-parameter fit implies shocks as the most singular dissipative structures. The β-parameter indicates a higher degree of intermittency in comparison to the Kolmogorov-Burgers model. The co-dimension corresponding to the best fit is $C \approx 1.23$, which now can be interpreted as a mixture of one-dimensional (vortex-like) and two-dimensional (sheet-like) structures. This is also suggested by the appearance of the structures associated with high vorticity in Fig. 3.2.

While both the one-parameter model with $C \approx 0.71$ ($\Delta = 2/3$) and the two-parameter model with $C \approx 1.23$ ($\Delta = 1$) agree very well with the exponents Z_1 and Z_2, the latter matches the higher-order statistics much better (see Fig. 3.7). Since the scaling exponents for $p > 3$ are subject to greater uncertainty, this might be accidental. However, the model with $\Delta = 1$ is more plausible for the following reason. By inverting Eq. (1.52) with $\beta = 1 - \Delta/C$ for $p = 2$, we can determine how the co-dimension varies with second-order exponent $Z_2 = 2\gamma/\zeta_3$, where $\gamma = \zeta_3/3 = 1/3$ for Kolmogorov scaling and $\gamma = \zeta_3/2$ for Burgers scaling. The

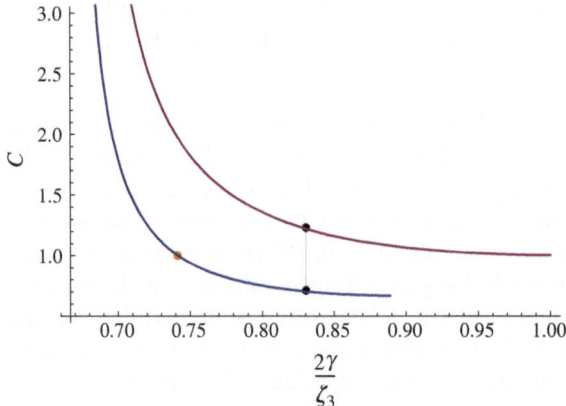

Fig. 3.8 Co-dimension $C = \Delta/(\beta - 1)$ as function of the second-order scaling index $Z_2 = 2\gamma/\zeta_3$ for the $\Delta = 2/3$ (*lower line*) and $\Delta = 1$ (*upper line*) [2]. The two dots connected by the vertical line correspond to the fits to the numerical results shown in Fig. 3.7. The Kolmogorov-Burgers model with $C = 1$ is marked by the orange dot

functions $C(2\gamma/\zeta_3)$ resulting for the two cases $\Delta = 2/3$ and $\Delta = 1$ are plotted in Fig. 3.8. For $\Delta = 2/3$, which includes the Kolmogorov-Burger model with $C(0.74) = 2$, no real solution exists if Z_2 becomes greater than about 0.89. In this case, the minimal co-dimension is slightly less than 0.7. The solution for $\Delta = 1$, on the other hand, converges to $C = 1$ for $Z_2 \to 1$, corresponding to $\gamma/\zeta_3 \to 0.5$. Thus, Burgers scaling follows from this model in the limit where dissipation is completely dominated by shocks.

To elaborate further on this conjecture, let us consider the simulations with purely solenoidal and compressive forcing, FLASH_SOLN and FLASH_COMP. Plots of $S_p^{\perp}$ versus $S_3^{\perp}$ averaged over the time interval $2 \leq t/T \leq 10$ are shown for the highest resolution ($N = 1024^3$) in Fig. 3.9. For a more reliable determination of the relative scaling exponents Z_p from power law fits, the fit range was adjusted such that the deviation of the data points from each fit function is less than 1 %, i. e.,

$$\forall p \leq 5 : \mathrm{err}_p := |\exp(\mathrm{fit}_p) - S_p^{\perp}|/S_p^{\perp} < 0.01,$$

where fit_p is a linear fit to $\log S_p^{\perp}$. The above criterion is fulfilled in the intervals $12.0 \leq S_3^{\perp} \leq 120$ for solenoidal forcing and $25.0 \leq S_3^{\perp} \leq 150$ for compressive forcing. The resulting scaling exponents Z_p are listed for all resolutions in Table 3.4. In the case $N = 1024^3$, the standard errors of the parameters Z_p are of the order 10^{-3}. Systematic errors due to the resolution limit and ambiguities in the fit range, however, could still be of the order 10^{-2}.

The time variation of the scaling exponents obtained from the instantaneous structure functions is shown in Fig. 3.10 for $N = 1024^3$, One can see that the exponents resulting from solenoidal forcing differ markedly from the case of compressive

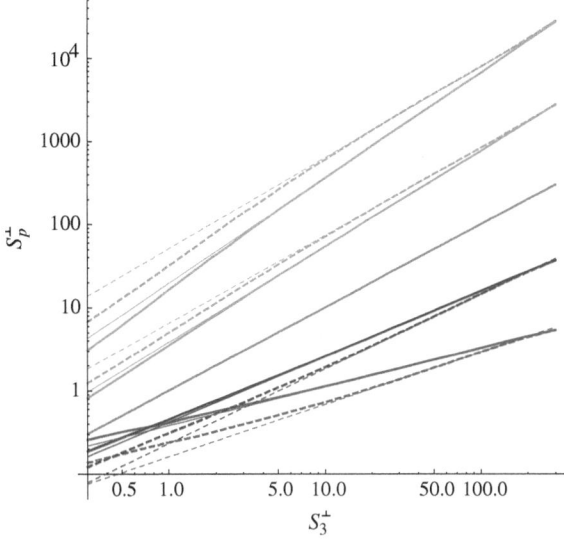

Fig. 3.9 Time-averaged structure functions $S_p^{\perp}$ versus $S_3^{\perp}$ for solenoidal (*solid lines*) and compressive (*dashed lines*) forcing [16]. The order p ranges form 1 (*bottom*) to 5 (*top*). The corresponding power-law fit functions are plotted as thin lines

Table 3.4 Relative scaling exponents Z_p from fits of time-averaged structure functions $S_p^{\perp}$ versus $S_3^{\perp}$

	N	Z_1	Z_2	Z_3	Z_4	Z_5
sol.	256^3	0.472	0.786	1.	1.160	1.289
	512^3	0.474	0.792	1.	1.149	1.265
	1024^3	0.466	0.788	1.	1.150	1.266
comp.	256^3	0.603	0.879	1.	1.072	1.126
	512^3	0.627	0.896	1.	1.056	1.097
	1024^3	0.628	0.897	1.	1.055	1.095

forcing. In each case, the variation of Z_1 and Z_2 clearly shows a temporal correlation, while the higher-order exponents appear to be anti-correlated. This suggests that the instantaneous scaling exponents show an imprint of the stochastic variation of the large scale forcing rather than purely statistical scatter. For incompressible turbulence, an influence of the large scales on much smaller scales is reported, for example, in [21, 22].

Fits of the Kolmogorov-Burgers model to the time-averaged scaling exponents show again large deviations both for solenoidal and compressive forcing (see Fig. 3.11). As for the simulation ENZO_HYBR, the fits imply co-dimensions below unity: $C_{sol} \approx 0.76$ for FLASH_SOLN and $C_{comp} \approx 0.67$ for FLASH_COMP, which is close to the minimum $C = \Delta = 2/3$. Although a good fit is obtained in the case of

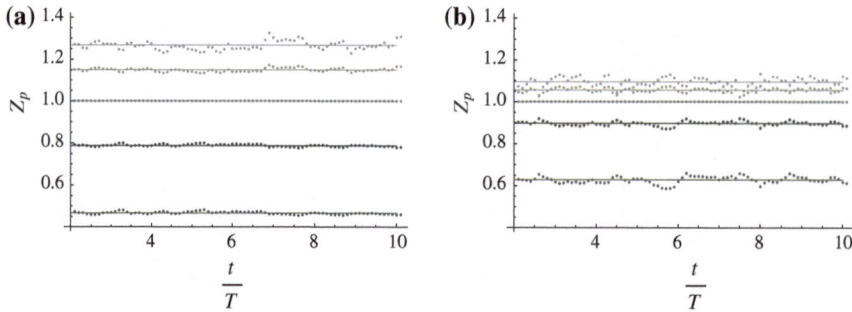

Fig. 3.10 Time evolution of the relative scaling exponents Z_p. The time averages are indicated by the horizontal lines [16]. **a** solenoidal **b** compressive

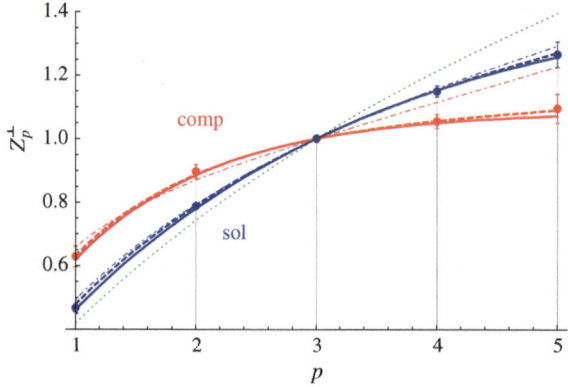

Fig. 3.11 Comparison of the relative scaling exponents Z_p for solenoidal and compressive forcing (see Table 3.4) with the Kolmogorov-Burgers model (*dotted line*) and fits of the general intermittency model (1.52) with $\Delta = 2/3$ (*dot-dashed lines*), $C = 1$ (*thick dashed lines*) and $\Delta = 1$ (*solid lines*) [16]. The vertical bars indicate the standard deviations of the instantaneous values

solenoidal forcing, the fit for compressive forcing is very poor. For $\Delta = 1$, on the other hand, fit functions with co-dimensions $C_{sol} \approx 1.5$ and $C_{comp} \approx 1.1$ agree very well with the data. In particular, the co-dimension is significantly closer to unity for compressively driven turbulence, corresponding to the stronger impact of shocks. This corroborates the conjecture that, in addition to the RMS Mach number as major parameter, the intermittency of the compressible turbulent cascade depends on the ratio of compressive and solenoidal modes of the large-scale energy injection.

As pointed out in Sect. 2.1, however, there is an alternative interpretation: The alteration of the intermittency parameters induced by the forcing might not reflect genuine inertial-range properties of the turbulent cascade, but distortions caused by extreme choices of the forcing. To settle this question, a more thorough understanding of the inertial-range dynamics of compressible turbulence is required. One possibility is the verification of analytical relations, such as Eq. (1.27), on numerical data [23, 24]. Another promising approach is spectral transfer analysis [25]. For example, this method was applied to compressible MHD turbulence in the solar convection zone to analyze the exchange between the kinetic and magnetic energy reservoirs on a scale-by-scale basis [27]. In a similar way, the calculation of transfer functions could be utilized to study the exchange between kinetic energy and pressure modes. This method also addresses the problem that a Kolmogorov-like inertial subrange cannot exist for turbulent velocity fluctuations in the compressible regime (see Sect. 1.2.2). Although two-point statistics of the velocity field exhibit power laws, the compressible cascade must also encompass pressure or—in the isothermal case—density fluctuations [26].

3.3 Mass-Weighted Two-Point Statistics

In [16], a reinterpretation of the intermittency model (1.52) in the context of compressible turbulence is proposed. Rather than considering the rate of energy dissipation per unit mass, the hierarchy of dissipative structures could be constructed from the volumetric dissipation rate $\tilde{\varepsilon}_l := \rho_l \varepsilon_l$, which is defined by the average of $\rho(\mathbf{x}, t)\varepsilon(\mathbf{x}, t)$ over a region of size l. This definition follows from the reasoning in Sect. 1.22 (see also [28]). For a consistent formulation, the refined similarity hypothesis should be expressed in the from

$$\tilde{S}_p(l) = \tilde{C}_p l^{\tilde{\zeta}_3 p/3} \langle (\rho_l \varepsilon_l)^{p/3} \rangle^{\tilde{\zeta}_3} \propto l^{\zeta_3(p/3 + \tau_p/3)}, \tag{3.8}$$

where $\tilde{S}_p(l) := \langle \delta \tilde{v}_l^p \rangle = \langle \delta(\rho^{1/3} v)_l^p \rangle$. Now, the multiplicative random process with log-Poisson statistics explained in Sect. 1.3 naturally leads to Eq. (1.52) for the relative scaling exponents of the mass-weighted velocity fluctuations,

$$\tilde{Z}_p = \frac{\tilde{\zeta}_p}{\tilde{\zeta}_3} = \frac{p}{3} + \tilde{\tau}_{p/3} . \tag{3.9}$$

In contrast to the original formulation, the third-order exponent $\tilde{\zeta}_3$ is not constrained to be unity. In [9], on the other hand, it is argued that linear scaling of the third-order structure function, $\tilde{S}_3(l) \propto \ell$, is the natural generalization of the incompressible four-fifth law (1.21), which expresses the constancy of the turbulence energy flux in the inertial range (see also [29] for MHD turbulence). This proposition is theoretically appealing, but it is not supported by all numerical simulations [1]. In the following, we consider only the relative scaling exponents.

Extended self-similarity plots of the mass-weighted structure functions $\tilde{S}_p^\perp(l)$ calculated from the FLASH_SOLN and FLASH_COMP data are shown in Fig. 3.12. Since the mass weighing introduces stronger fluctuations, the fit functions satisfy the error criterion defined in Sect. 3.2 only in the narrower ranges $15.0 \leq \tilde{S}_3^\perp \leq 100$ and $25.0 \leq \tilde{S}_3^\perp \leq 75.0$ for solenoidal and compressive forcing, respectively.

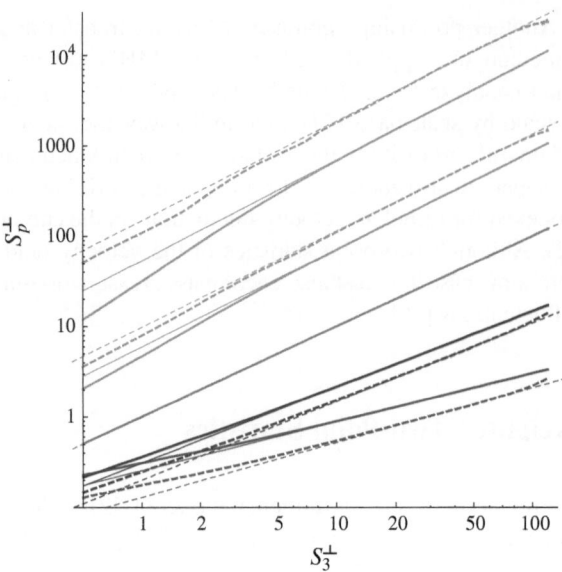

Fig. 3.12 Time-averaged structure functions $\tilde{S}_p^\perp$ versus $\tilde{S}_3^\perp$ of the mass-weighted velocity $\tilde{\mathbf{v}} = \rho^{1/3}\mathbf{v}$, as in Fig. 3.9 [16]

Table 3.5 Relative scaling exponents $\tilde{Z}_p$ from fits of time-averaged mass-weighted structure functions $\tilde{S}_p^\perp$ versus $\tilde{S}_3^\perp$ (see Fig. 3.12)

	N	$\tilde{Z}_1$	$\tilde{Z}_2$	$\tilde{Z}_3$	$\tilde{Z}_4$	$\tilde{Z}_5$
Sol.	256^3	0.546	0.839	1.	1.094	1.150
	512^3	0.550	0.845	1.	1.082	1.122
	1024^3	0.539	0.840	1.	1.080	1.112
Comp.	256^3	0.635	0.893	1.	1.034	1.026
	512^3	0.634	0.887	1.	1.050	1.068
	1024^3	0.605	0.869	1.	1.066	1.100

The resulting relative scaling exponents $\tilde{Z}_p$ are summarized in Table 3.5. Although the sensitivity on resolution is more pronounced than for the pure velocity exponents Z_p and the scatter is much larger, it appears that the differences between the two different types of forcing are substantially reduced.

As shown in Fig. 3.13, neither the Kolmogorov-Burgers model nor the best-fit model with $\Delta = 2/3$ match the scaling exponents $\tilde{Z}_p$. The closest fits are obtained for $\Delta = 1$, where $C_{\text{sol}} \approx 1.18$ and $C_{\text{comp}} \approx 1.08$. Fits with fixed co-dimension $C = 1$ yield $\Delta_{\text{sol}} = 0.90$ and $\Delta_{\text{comp}} \approx 0.92$, which is closer to unity than to the Kolmogorov exponent $2/3$. In conclusion, the same trends can be seen as for the pure velocity scaling exponents in Sect. 3.2, although the difference between the co-dimensions is smaller and the lower values point at a stronger influence of shocks.

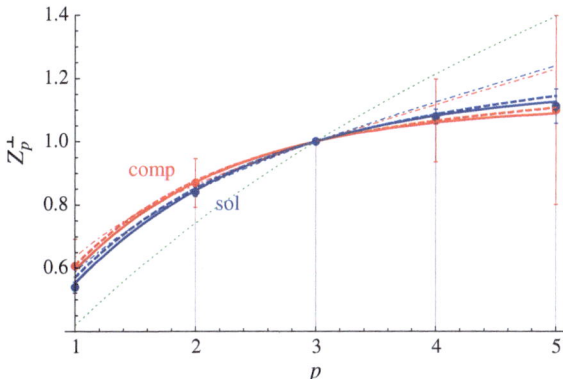

Fig. 3.13 Relative scaling exponents $\tilde{Z}_p$ of the mass-weighted velocity $\tilde{\mathbf{v}} = \rho^{1/3}\mathbf{v}$ for solenoidal and compressive forcing and various fit functions, as in Fig. 3.11 [16]

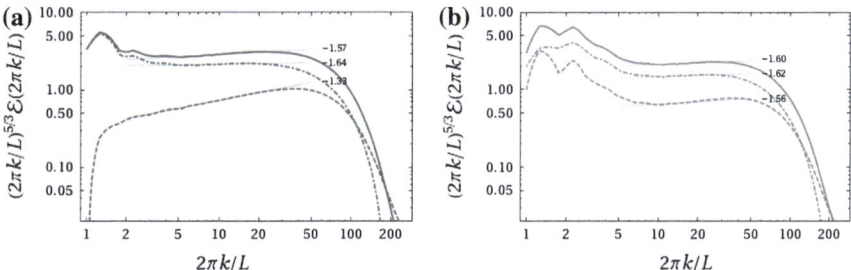

Fig. 3.14 Kinetic energy spectrum functions (Eq. 1.37) for solenoidal and compressive forcing compensated with $k^{5/3}$. Transversal and longitudinal spectra are plotted as dot-dashed and dashed lines, respectively. The thin lines show power-law fits **a** solenoidal **b** compressive

A complementary view is provided by the energy spectra defined by Eq. (1.37). Figure 3.14 shows $(2\pi k/L)^{5/3}\mathscr{E}(k)$ for FLASH_SOLN and FLASH_COMP. A spectrum with slope $-5/3$ would appear as a horizontal line in this representation. Although the spectrum of turbulence produced by solenoidal forcing is slightly flatter, the transversal part (projection of the Fourier modes $\widehat{\mathbf{v}}(\mathbf{k})$ and $\widehat{\rho\mathbf{v}}(\mathbf{k})$ perpendicular to the wave vector $\mathbf{k}$) exhibits a slope very close to $-5/3$. This is similar to the results for the spectrum of $\tilde{\mathbf{v}}^2$ reported in [1, 7, 9]. For compressive forcing, on the other hand, the spectra are much steeper for a relatively narrow range of wave numbers below the forcing wave number. In [7], this behavior is confirmed by computing the spectrum of $\tilde{\mathbf{v}}^2$ for compressively driven turbulence with higher RMS Mach number. In this case, the slope of about -2.1 is consistent with theoretical predictions [30]. For k greater than about 10, the right plot in Fig. 3.14 shows that $\mathscr{E}(k)$ flattens to a slope slightly above $-5/3$. Given the limited numerical resolution, this simply might be a consequence of the bottleneck effect, a bump-like distortion of the spectrum near the range of wave numbers affected by numerical dissipation [9, 31–33]. However, it could also hint at a genuine flattening of the spectrum that restores universal scaling on sufficiently small length scales, as proposed in [34].

3.4 Subgrid Scale Statistics

The data from the high-resolution simulations FLASH_SOLN and FLASH_COMP allow for the explicit computation of the turbulence energy flux across a length scale that is both large compared to the grid resolution and small compared to the length scale of energy injection by the forcing. This can be used to test subgrid scale models [35] for LES. The statistics of the subgrid scale turbulence energy in LES with different resolutions has further implications on the scaling of supersonic turbulence.

3.4.1 Turbulence Energy Flux

The turbulence energy flux across a given length scale Δ_G is explicitly defined by

$$\Sigma_{\Delta_G} = \left(-\overline{\rho v_i v_j} + \frac{\overline{\rho v_i}\,\overline{\rho v_i}}{\overline{\rho}} \right) \overline{v}_{i,j} \,, \tag{3.10}$$

where the first factor on the right hand side is the turbulence stress tensor defined analogous to Eq. (2.21) and the second factor is the derivative of the filtered velocity. As a shorthand notation, filtering is indicated by an overline, for instance, $\overline{\rho} = \langle \rho \rangle_{\Delta_G}$, and $\overline{v}_i = \langle \rho v_i \rangle_{\Delta_G} / \overline{\rho}$, where the unfiltered dynamical variables are assumed to be computed with implicit large eddy simulations (ILES), such as FLASH_SOLN and FLASH_COMP. In the following, we assume that the grid scale Δ of the ILES is small compared to filter length Δ_G.

A commonly used approximation of Σ_{Δ_G} in terms of filtered variables is the eddy-viscosity closure:

$$\Sigma_{\Delta_G}^{(\mathrm{cls})} = C_1 \Delta_G (2\overline{\rho} K_{\Delta_G})^{1/2} |\overline{S}^*|^2 - \frac{2}{3} K_{\Delta_G} \overline{d}, \tag{3.11}$$

where

$$\overline{S}_{ij}^* = \frac{1}{2} \left(\frac{\partial \overline{v}_i}{\partial x_j} + \frac{\partial \overline{v}_j}{\partial x_i} \right) - \frac{1}{3} \overline{d} \delta_{ij} \tag{3.12}$$

is the trace-free rate-of-strain tensor and $\overline{d}$ the divergence of the filtered velocity field. The turbulence energy on length scales smaller than Δ_G is defined by

$$K_{\Delta_G} = \frac{1}{2} \left[\overline{\rho v^2} - \frac{(\overline{\rho v})^2}{\overline{\rho}} \right]. \tag{3.13}$$

Strictly speaking, K_{Δ_G} is the turbulence energy for length scales ranging from the grid resolution Δ to the smoothing length Δ_G. However, this distinction can be neglected if $\Delta \ll \Delta_G$ [36].

If Σ_{Δ_G} is know, the coefficient C_1 can be determined by calculating the least-square match of $\Sigma_{\Delta_G}^{(\text{cls})}$ to Σ_{Δ_G}. By defining

$$C_1 f^{(\text{cls})} = \Sigma_{\Delta_G}^{(\text{cls})} + \frac{2}{3} K_{\Delta_G} \overline{d}, \tag{3.14}$$

where $f^{(\text{cls})} = \Delta_G (2\overline{\rho} K_{\Delta_G})^{1/2} |\overline{S}^*|^2$, the squared error function of the closure can be written as

$$\text{err}^2(C_1) = \int_{\mathcal{V}} \left| \Sigma_{\Delta_G} + \frac{2}{3} K_{\Delta_G} \overline{d} - C_1 f^{(\text{cls})} \right|^2 d^3 x, \tag{3.15}$$

where Σ_{Δ_G} and K_{Δ_G} are given by Eqs. (3.10) and (3.13), respectively, and the volume integral extends over the whole domain $\mathcal{V}$. The minimum of $\text{err}^2(C_1)$ yields the least squares error solution for the closure coefficient [35],

$$C_1 = \frac{\int_{\mathcal{V}} f^{(\text{cls})} \left[\Sigma_{\Delta_G} + \frac{2}{3} K_{\Delta_G} \overline{d} \right] d^3 x}{\int_{\mathcal{V}} |f^{(\text{cls})}|^2 d^3 x}. \tag{3.16}$$

For statistically stationary and isotropic turbulence, the mean energy flux is constant in the inertial subrange. Consequently, C_1 should be independent of the filter length scale and can be applied in LES with different resolutions.

In the following, we use the FLASH_SOLN and FLASH_COMP data to determine C_1. The turbulence energy flux is computed from Eq. (3.10), where the density and velocity fields are smoothed over the length scale $\Delta_G = 32\Delta$ with a Gaussian filter. The flux predicted by the eddy-viscosity closure is given by Eq. (3.11). The correlation between $\Sigma_{\Delta_G}^{(\text{cls})}$ and Σ_{Δ_G} is shown in Figs. 3.15 and 3.16. Although the correlation is quite good (the spacing of the contour lines is logarithmic), there is clearly a problem with negative fluxes, which correspond to the so-called backscattering effect in the turbulent cascade. Apart from that, the positive fluxes are somewhat tilted to lower values. The values of the closure coefficient C_1 following from Eq. (3.16) are listed in Table 3.6. Also listed are the correlation coefficients

$$\text{corr}\left[\Sigma_{\Delta_G}, \Sigma_{\Delta_G}^{(\text{cls})} \right] = \frac{\int_{\mathcal{V}} \left[\Sigma_{\Delta_G} - \langle \Sigma_{\Delta_G} \rangle \right] \left[\Sigma_{\Delta_G}^{(\text{cls})} - \langle \Sigma_{\Delta_G}^{(\text{cls})} \rangle \right] d^3 x}{\text{std}[\Sigma_{\Delta_G}] \, \text{std}\left[\Sigma_{\Delta_G}^{(\text{cls})} \right]}, \tag{3.17}$$

where $\text{std}[\cdot]$ denotes the standard deviation and the angle brackets indicate an average over the whole domain.

Since the eddy-viscosity closure was devised for incompressible turbulence, the question of whether shocks can be accommodated in closures for the turbulence energy flux has to be investigated. One idea is to simply set Σ equal to zero in the

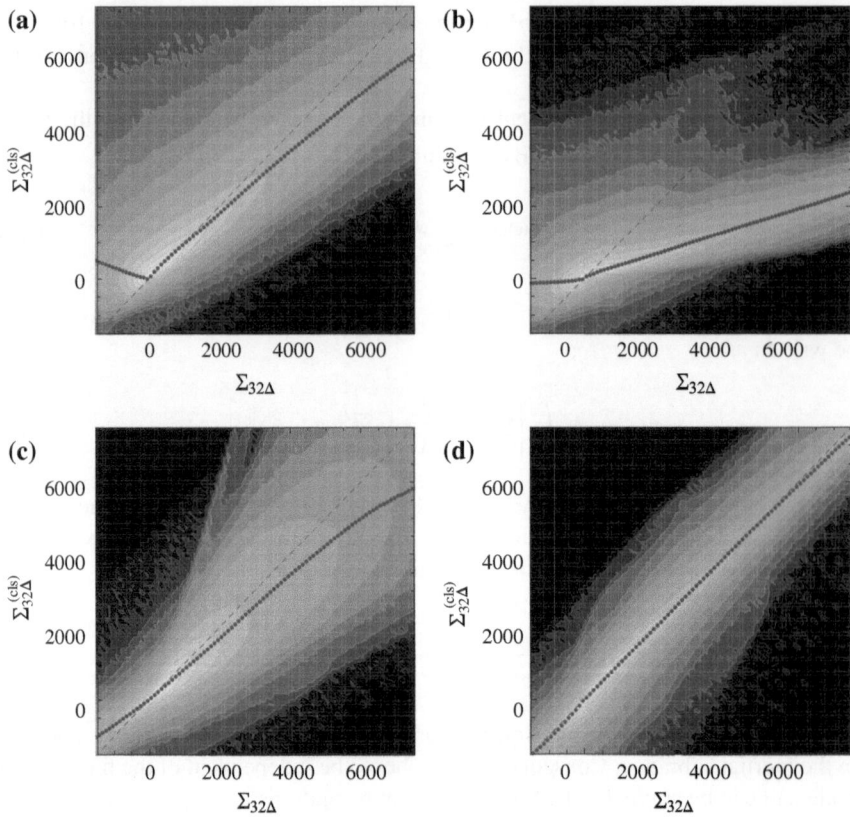

Fig. 3.15 Correlation diagrams for the SGS turbulence energy flux Σ defined by Eq. (3.10) in the case of isothermal supersonic turbulence with solenoidal forcing ($\mathcal{M}_{\mathrm{rms}} \approx 5.3$) [35]. The applied filter length is 32Δ. The blue dots indicate the average prediction of the closure for a given value of $\Sigma_{32\Delta}$ **a** eddy-viscosity closure **b** shocks excluded **c** determinant closure **d** non-linear closure

vicinity of shock fronts [37]. This should suppress overproduction of SGS turbulence energy by the very large strain at shock fronts. However, panels (b) in Figs. 3.15 and 3.16 make clear that excluding shocks substantially deteriorates the correlations and causes a significant underestimate of large positive fluxes. Although the applied shock detection criterion $\overline{d} < -\overline{c}_s/\Delta_G$ is rather crude, the trends indicate that shocks must not be separated from the supersonic turbulent cascade.

A different variant of the eddy-viscosity closure is based on the determinant of the velocity gradient. This closure was tested for compressible turbulence in the transonic regime [38]. Although the trace-free part of the SGS turbulence stress tensor is still given by the expression $\tau_{ij}^* = 2\rho \nu_{\mathrm{sgs}} S_{ij}^*$, the eddy viscosity does not depend on K_{sgs}. It is defined by the determinant of the trace-free rate-of-strain tensor:

$$\nu_{\mathrm{sgs}} = -C_1 \Delta^2 |S^*|^{-2} \det \mathbf{S}^*.$$

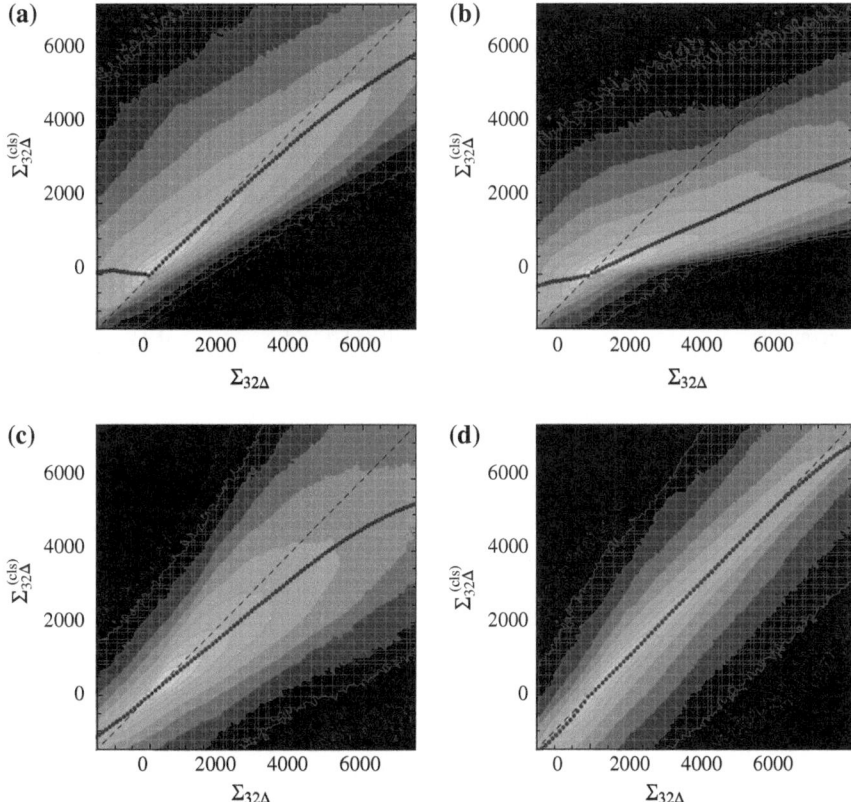

Fig. 3.16 Correlation diagrams for isothermal supersonic turbulence with compressive forcing ($\mathcal{M}_{\mathrm{rms}} \approx 5.6$), as in Fig. 3.15 [35]. **a** eddy-viscosity closure **b** shocks excluded **c** determinant closure **d** non-linear closure

Hence, the turbulence energy flux on the filter scale is given by

$$\Sigma_{\Delta_G}^{(\mathrm{cls})} = -C_1 \overline{\rho} \Delta_G^2 \det \overline{\mathbf{S}}^* - \frac{2}{3} K_{\Delta_G} \overline{d}. \tag{3.18}$$

A particularly interesting feature of the determinant is that it switches signs and thereby describes two different flow topologies [38]. In one case, the determinant is negative. This corresponds to the forward turbulent cascade transporting energy from large eddies to smaller eddies. In the other case, the flow is contracting in one dimension and expanding in the others. Then the determinant is positive, corresponding to a backscattering of energy from small eddies to larger eddies. This phenomenon can be explained by the alignment of vortices along a single stretching direction (the "tornado" topology). While an energy flux of the form (3.11) fails to describe the reverse cascade, one can see in Figs. 3.15c and 3.16c that the determinant closure

Table 3.6 Closure and correlation coefficients for the closures shown in Figs. 3.15 and 3.16

Closure	C_1	Corr $\left[\Sigma_{32\Delta}, \Sigma_{32\Delta}^{(cls)} \right]$
Solenoidal forcing, $\zeta = 1.0$, $\mathscr{M}_{rms} \approx 5.3$		
Eddy viscosity	0.102	0.950
Eddy viscosity (shocks excluded)	0.055	0.931
Determinant	0.803	0.950
Non-linear	0.849	0.991
Compressive forcing, $\zeta = 0.0$, , $\mathscr{M}_{rms} \approx 5.6$		
Eddy viscosity	0.092	0.930
Eddy viscosity (shocks excluded)	0.059	0.914
Determinant	0.834	0.947
Non-linear	0.833	0.991

yields a good correlation to the negative energy flux. However, the overall correlation does not significantly improve (see Table 3.6), because of the relatively large scatter in the forward cascade.

In [39], a non-linear expression for the turbulence stress tensor is investigated, which depends on the full Jacobian matrix of the velocity:

$$\tau_{ij} = -2C_1 K_{sgs} \frac{2v_{i,k}v_{j,k}}{|\nabla \otimes \mathbf{v}|^2} - \frac{2}{3}(1 - C_1)K_{sgs}\delta_{ij}. \tag{3.19}$$

Since $|\nabla \otimes \mathbf{v}| = (2v_{i,k}v_{i,k})^{1/2}$, the above expression fulfills the identity $\tau_{ii} = -2K_{sgs}$. The corresponding turbulence energy flux on the filter length scale Δ_G is given by

$$\Sigma_{\Delta_G}^{(cls)} = -4C_1 K_{\Delta_G} \frac{\overline{v}_{i,k}\overline{v}_{j,k}\overline{S}_{ij}^*}{|\nabla \otimes \overline{\mathbf{v}}|^2} - \frac{2}{3}(1 - C_1)K_{\Delta_G}\overline{d}. \tag{3.20}$$

Figs. 3.15d and 3.16d show that the correlation is excellent for the above closure, with correlation coefficients above 0.99 (see Table 3.6). Like the determinant closure discussed above, the trace-free part of the non-linear closure for the SGS turbulence stress switches signs and, thus, allows for a backward energy cascade.

To compare the quality of the different closures, slices of the two-dimensional probability density functions are plotted in Fig. 3.17. While both the determinant and the non-linear closures yield good approximations to negative values of the energy flux, the non-linear closure is clearly superior for large positive fluxes. However, the tails are flatter for compressively driven turbulence, which implies that large deviations are more frequent in this case. Relatively small positive energy flux is also well reproduced by the standard eddy-viscosity closure. In [35], linear combinations of different closures are investigated. The main result is that a combination of the eddy-viscosity closure with the coefficient $C_1 = 0.02$ and the non-linear closure with the coefficient $C_2 = 0.7$ (see Eq. 2.26) has optimal properties for LES of supersonic turbulence.

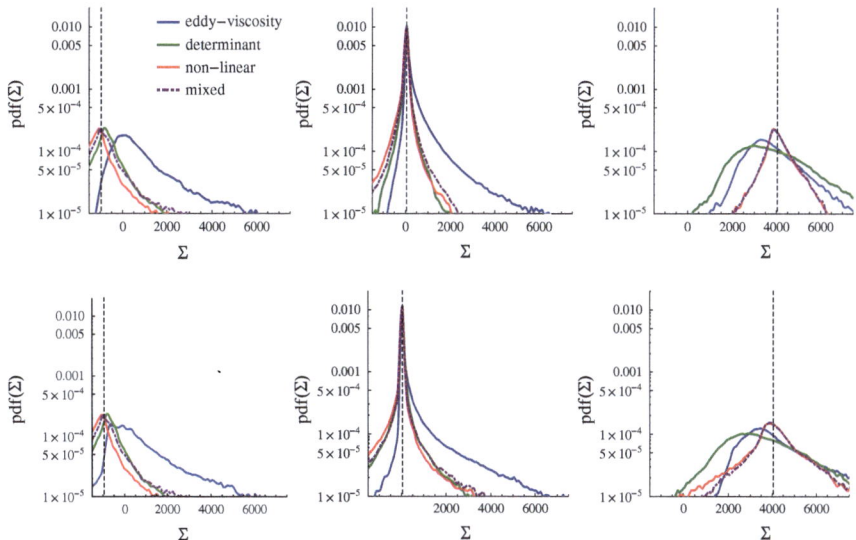

Fig. 3.17 Slices of the two-dimensional probability density functions plotted in Figs. 3.15 and 3.16, showing the predictions of different closures for the values of the explicitly computed energy flux $\Sigma_{32\Delta}$ that are indicated by the vertical dashed lines. The top and bottom rows of panels show the results for solenoidal and compressive forcing, respectively [35]

3.4.2 Scaling of the Subgrid Scale Turbulence Energy

By performing LES, as explained in Sect. 2.3, the scaling of turbulent velocity fluctuations can be inferred in a way that differs from both structure functions and turbulence energy spectra. Based on the scale-decomposition of the equations of fluid dynamics for a cutoff length Δ, the energy that is associated with fluctuations of the velocity and density fields on length scales $\ell \lesssim \Delta$ is defined by expression (2.22). In LES, the partial differential Eq. (2.25) is solved for $K_{\text{sgs}}(\mathbf{x}, t)$. The scaling behavior of the mean unresolved energy fraction, $\langle K_{\text{sgs}} \rangle$, can be determined by varying the grid resolution Δ. In the following, we analyze LES with Δ ranging from $L/256$ to $L/32$ [35]. The case $\Delta = L/256$ corresponds to the simulations presented in Sect 2.3.

Spatial averages of the RMS Mach number and the SGS turbulence Mach number are plotted in Fig. 3.18 (top panels) as functions of time for the different LES. The flow approaches a statistically stationary state after about 2 integral time scales T, for which $\mathcal{M}_{\text{rms}}$ saturates around values between 5 and 6. The RMS Mach number of the turbulent velocity fluctuations below the grid scale is given by $\langle \mathcal{M}_{\text{sgs}}^2 \rangle^{1/2}$, where $\mathcal{M}_{\text{sgs}}$ is defined by Eq. (2.24). As one can see in the bottom panels of Fig. 3.18, $\langle \mathcal{M}_{\text{sgs}}^2 \rangle^{1/2}$ decreases with the grid scale Δ. This reflects the decrease of the unresolved energy fraction, as the cutoff of the numerically resolved spectrum is shifted to smaller length scales. By averaging the RMS Mach numbers from $t = 2T$ to $10T$, the time-averaged mean values listed in Table 3.7 are obtained. The time averages of $\langle \mathcal{M}_{\text{sgs}}^2 \rangle^{1/2}$ closely

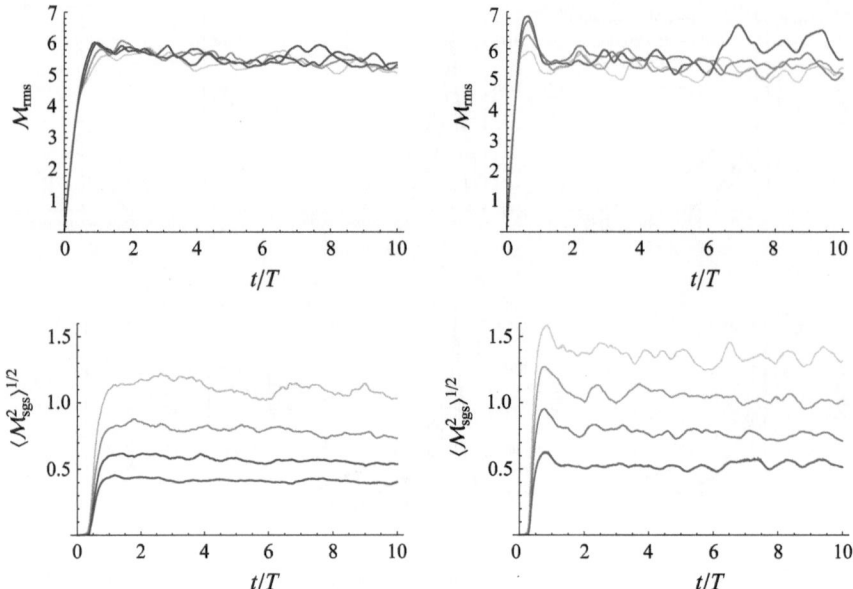

Fig. 3.18 Temporal evolution of the RMS Mach number (*top*) and the mean SGS turbulence Mach number (*bottom*) for solenoidal (*left column*) and compressive forcing (*right column*) [35]. The cutoff length Δ decreases from $L/32$ (*light colour*) to $L/256$ (*full colour*)

Table 3.7 Time-averaged spatial mean values of various quantities and their standard deviations from the averages for different numerical resolutions

N	Δ/L	$\mathcal{M}_{rms}$	$\langle\mathcal{M}_{sgs}^2\rangle^{1/2}$	$\langle K_{sgs}\rangle/(\rho_0 V^2)$	$(L/V^3)\langle\varepsilon\rangle$
Solenoidal forcing ($\zeta = 1$)					
64	1/32	5.38	1.107 ± 0.053	0.0726 ± 0.0055	1.236 ± 0.141
128	1/64	5.50	0.787 ± 0.030	0.0407 ± 0.0025	1.230 ± 0.113
256	1/128	5.55	0.578 ± 0.022	0.0236 ± 0.0013	1.213 ± 0.098
512	1/256	5.52	0.412 ± 0.012	0.0138 ± 0.0012	1.219 ± 0.159
Compressive forcing ($\zeta = 0$)					
64	1/32	5.29	1.353 ± 0.049	0.0235 ± 0.0044	0.253 ± 0.072
128	1/64	5.43	1.040 ± 0.041	0.0148 ± 0.0018	0.286 ± 0.047
256	1/128	5.57	0.767 ± 0.029	0.0086 ± 0.0013	0.293 ± 0.061
512	1/256	5.86	0.528 ± 0.023	0.0048 ± 0.0008	0.292 ± 0.067

follow power laws, as shown in Fig. 3.19 a:

$$\langle\mathcal{M}_{sgs}^2\rangle^{1/2} \propto \Delta^{\alpha_1}. \tag{3.21}$$

The slopes are $\alpha_1 = 0.475 \pm 0.004$ for solenoidal and 0.451 ± 0.026 for compressive forcing.

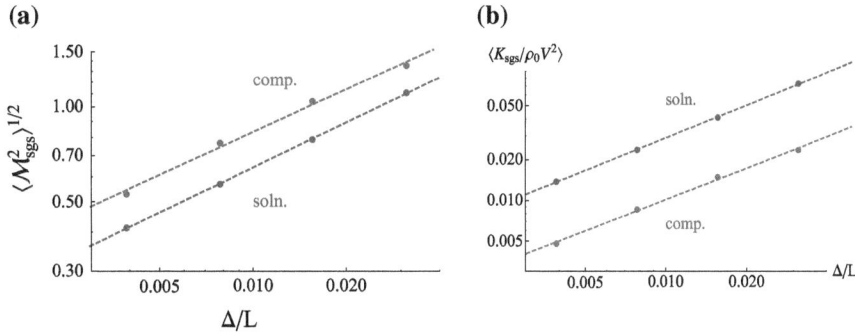

Fig. 3.19 Scaling laws for the mean SGS turbulence Mach number (**a**) and energy (**b**) as functions of the numerical resolution Δ [35].

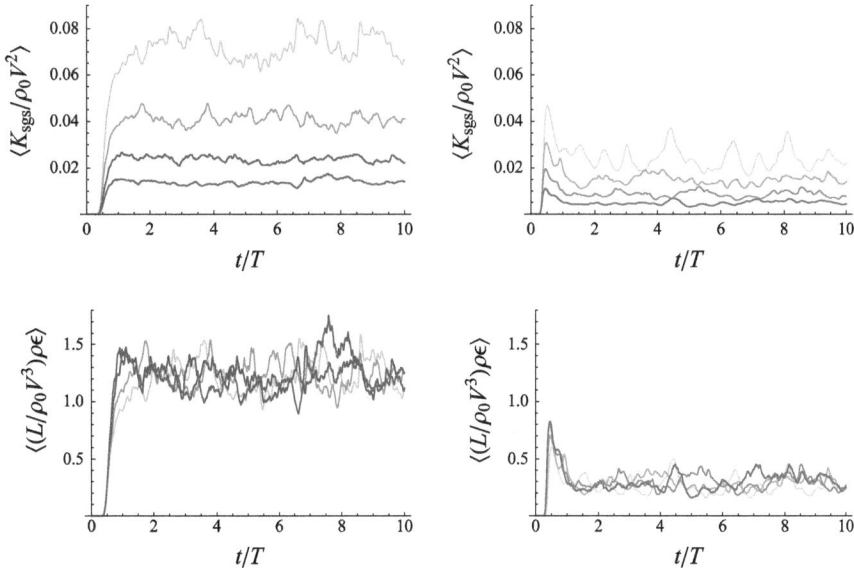

Fig. 3.20 Temporal evolution of the SGS turbulence energy (*top*) and the dissipation rate (*bottom*) for solenoidal (*left column*) and compressive forcing (*right column*) [35]. The cutoff length Δ decreases from $L/32$ (*light colour*) to $L/256$ (*full colour*)

The behavior of the mean SGS turbulence energy is similar (see Fig. 3.20), although the fluctuations of $\langle K_{\mathrm{sgs}} \rangle$ are more pronounced in comparison to $\langle \mathcal{M}_{\mathrm{sgs}}^2 \rangle^{1/2}$. This is caused by the large fluctuations of the mass density, which is included in K_{sgs}. For compressive forcing, $\langle K_{\mathrm{sgs}} \rangle$ is systematically lower in comparison to solenoidal forcing. Thus, the total amount of energy in the turbulent structures on a given length scale is smaller in the case of compressive forcing, as suggested by the plots of K_{sgs} in Figs. 2.7 and 2.8. On the other hand, the scaling law

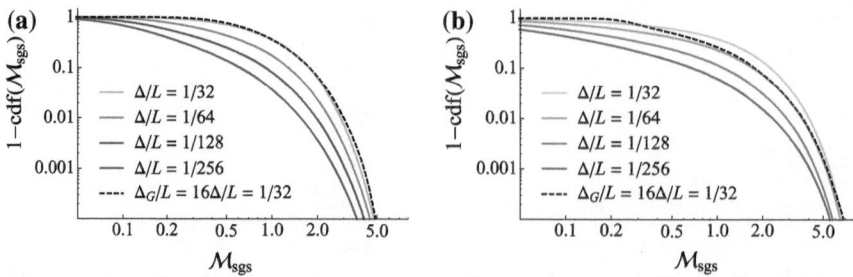

Fig. 3.21 Volume fractions of grid cells in which the SGS turbulence Mach number is greater than a given value for different numerical resolutions [35]. The dashed lines follows from the filtering of 1024^3 ILES data with filter length 16Δ. This corresponds to the 64^3 LES, for which $\Delta/L = 1/32$ **a** solenoidal **b** compressive

$$\langle K_{\text{sgs}} \rangle \propto \Delta^{\alpha_2}, \tag{3.22}$$

has nearly the same exponents for solenodial and compressive forcing (see Fig. 3.19 b). Power-law fits yield $\alpha_2 = 0.799 \pm 0.009$ and 0.769 ± 0.029, which agree within the error bars. This result is remarkable because it indicates that the scaling properties of turbulence on small length scales are independent of the forcing, similar to the scalings inferred from the mass-weighted structure functions and energy spectra in Sect. 3.3. The mean dissipation rate $\langle \rho\varepsilon \rangle$ is plotted in the bottom panels of Fig. 3.20. The time-averaged mean values listed in Table 3.7 demonstrate that $\langle \rho\varepsilon \rangle$ is independent of the cutoff scale Δ, which is an essential property of turbulent energy dissipation (see Sect. 1.2.1).

The contribution of SGS turbulence to the effective pressure (Eq. 2.23) is large if $\mathcal{M}_{\text{sgs}} \gtrsim 1$. In Fig. 3.21, $[1 - \text{cdf}(\mathcal{M}_{\text{sgs}})]$ is plotted, where $\text{cdf}(\mathcal{M}_{\text{sgs}})$ is the cumulative distribution function of $\mathcal{M}_{\text{sgs}}$. The graphs specify the volume fraction of cells in which the SGS turbulence Mach number is greater than a given value. As expected, the fraction of cells for which $\mathcal{M}_{\text{sgs}} > 1$ decreases with the grid scale Δ. However, the high-Mach tails indicate that supersonic velocity fluctuations occur on unresolved length scales even at relatively high resolution. For the lowest-resolution LES with the grid scale $\Delta = L/32$, we can compare the distribution of $\mathcal{M}_{\text{sgs}}$ to the distribution that is obtained by smoothing the 1024^3 ILES data over the same length scale $\Delta_{\text{G}} = 16\Delta_{\text{ILES}} = L/32$ (see Sect. 3.4.1). As shown in Fig. 3.21, the distributions agree remarkably well in the case of solenoidal forcing. For compressive forcing, there are larger discrepancies. Since Gaussian filtering corresponds only roughly to the implicit filter of LES and the SGS model is based on various approximations, the match is nevertheless satisfactory. The larger deviations in the case of compressive forcing point toward a missing physical effect such as the pressure-dilatation, which is entirely neglected in the applied SGS model.

References

1. C. Federrath, J. Roman-Duval, R.S. Klessen, W. Schmidt, M. Mac Low, A&A 512, A81+ (2010). doi:10.1051/0004-6361/200912437
2. W. Schmidt, C. Federrath, M. Hupp, S. Kern, J.C. Niemeyer, A&A **494**, 127 (2009). doi:10. 1051/0004-6361:200809967
3. B. Fryxell, K. Olson, P. Ricker, F.X. Timmes, M. Zingale, D.Q. Lamb, P. MacNeice, R. Rosner, J.W. Truran, H. Tufo, ApJS **131**, 273 (2000). doi:10.1086/317361
4. B.W. O'Shea, G. Bryan, J. Bordner, M.L. Norman, T. Abel, R. Harkness, A. Kritsuk, in *Adaptive Mesh Refinement—Theory and Applications*, ed. by T. Plewa, T. Linde, V.G. Weirs, Lecture Notes in Computational Science and Engineering, vol. 41, p. 341. http://esoads.eso.org/abs/ 2004astro.ph.3044O
5. U. Frisch, *Turbulence. The legacy of A.N. Kolmogorov* (Cambridge University Press, Cambridge, 1995)
6. S.B. Pope, *Turbulent Flows* (Cambridge University Press, Cambridge, 2000)
7. C. Federrath, *On the universality of compressible supersonic turbulence* (2013) Submitted to MNRAS
8. S. Kida, S.A. Orszag, J. Sci. Comput. **5**, 85 (1990)
9. A.G. Kritsuk, M.L. Norman, P. Padoan, R. Wagner, ApJ **665**, 416 (2007). doi:10.1086/519443
10. E. Falgarone, J. Puget, M. Perault, A&A **257**, 715 (1992)
11. M.S. Miesch, J. Bally, ApJ **429**, 645 (1994). doi:10.1086/174352
12. P. Padoan, M. Juvela, A. Kritsuk, M.L. Norman, ApJ **653**, L125 (2006). doi:10.1086/510620
13. C.M. Brunt, M.H. Heyer, ApJ **566**, 289 (2002). doi:10.1086/338032
14. R. Benzi, S. Ciliberto, R. Tripiccione, C. Baudet, F. Massaioli, S. Succi, Phys. Rev. E **48**, 29 (1993). doi:10.1103/PhysRevE.48.R29
15. Z.S. She, E. Leveque, Phys. Rev. Lett. **72**, 336 (1994). doi:10.1103/PhysRevLett.72.336
16. W. Schmidt, C. Federrath, R. Klessen, Phys. Rev. Lett. **101**(19), 194505 (2008). doi:10.1103/ PhysRevLett.101.194505
17. A.G. Kritsuk, P. Padoan, R. Wagner, M.L. Norman, in *Turbulence and Nonlinear Processes in Astrophysical Plasmas, American Institute of Physics Conference Series*, ed. by D. Shaikh, G.P. Zank, vol. 932, pp. 393–399 (2007). doi:10.1063/1.2778991
18. S. Boldyrev, ApJ **569**, 841 (2002)
19. P. Padoan, R. Jimenez, A. Nordlund, S. Boldyrev, Phys. Rev. Lett. **92**(19), 191102 (2004). doi:10.1103/PhysRevLett.92.191102
20. J.M. Burgers, Adv. Appl. Mech. **1**, 171 (1948)
21. A. Alexakis, P.D. Mininni, A. Pouquet, Phys. Rev. Lett. **95**(26), 264503 (2005). doi:10.1103/ PhysRevLett.95.264503
22. C.M. Casciola, P. Gualtieri, B. Jacob, R. Piva, Phys. Fluids **19**, 1704 (2007). doi:10.1063/1. 2800043
23. R. Wagner, G. Falkovich, A.G. Kritsuk, M.L. Norman, J. Fluid Mech. **713**, 482 (2012). doi:10. 1017/jfm.2012.470
24. Kritsuk, Wagner, Norman, J. Fluid Mech. **729**, R1 (2013). doi:10.1017/jfm.2013.342
25. R.H. Kraichnan, J. Fluid Mech. **47**, 525 (1971). doi:10.1017/S0022112071001216
26. Aluie, Physica D **247**(1), 54–65 (2013). doi:10.1016/j.physd.2012.12.009
27. J. Pietarila Graham, R. Cameron, M. Schüssler, ApJ **714**, 1606 (2010). doi:10.1088/0004-637X/714/2/1606
28. R.C. Fleck Jr, ApJ **458**, 739 (1996). doi:10.1086/176853
29. A.G. Kritsuk, S.D. Ustyugov, M.L. Norman, P. Padoan, J.Phys. Conf. Ser. **180**(1), 012020 (2009). doi:10.1088/1742-6596/180/1/012020
30. S. Galtier, S. Banerjee, Phys. Rev. Lett. **107**(13), 134501 (2011). doi:10.1103/PhysRevLett. 107.134501
31. G. Falkovich, Phys. Fluids **6**, 1411 (1994). doi:10.1063/1.868255
32. W. Dobler, N.E. Haugen, T.A. Yousef, A. Brandenburg, Phys. Rev. E. **68**(2), 26304 (2003). doi:10.1103/PhysRevE.68.026304

33. W. Schmidt, W. Hillebrandt, J.C. Niemeyer, Comp. Fluids. **35**, 353 (2006)
34. A.G. Kritsuk, S.D. Ustyugov, M.L. Norman, P. Padoan, in numerical modeling of space plasma flows, Astronum-2009, vol. 429, ed. by N.V. Pogorelov, E. Audit, G.P. Zank Astronomical Society of the Pacific Conference Series, (2010), p. 15
35. W. Schmidt, C. Federrath, A&A 528, A106+ (2011). doi:10.1051/0004-6361/201015630.
36. W. Schmidt, J.C. Niemeyer, W. Hillebrandt, A&A **450**, 265 (2006). doi:10.1051/0004-6361: 20053617
37. A. Maier, L. Iapichino, W. Schmidt, J.C. Niemeyer, ApJ **707**, 40 (2009). doi:10.1088/0004-637X/707/1/40
38. P.R. Woodward, D.H. Porter, I. Sytine, S.E. Anderson, A.A. Mirin, B.C. Curtis, R.H. Cohen, W.P. Dannevik, A.M. Dimits, D.E. Eliason, K.H. Winkler, S.W. Hodson, in, *Proceedings of the Fourth UNAM Supercomputing Conference on Computational Fluid Dynamics* June 2000, ed. by E. Ramos, G. Cisneros, R. Fernandez-Flores, A. Santillan-Gonzalez (World Scientific, Mexico, 2001), pp. 3–15
39. P.R. Woodward, D.H. Porter, S. Anderson, T. Fuchs, F. Herwig, J.Phys. Conf. Ser. **46**, 370 (2006). doi:10.1088/1742-6596/46/1/052

Chapter 4
Turbulent Density Statistics

The statistics of density fluctuations play a central role for the phenomenology of compressible turbulence, particularly in the supersonic regime. The most important property is the probability density function of the mass density, which is often approximated by a log-normal function. Data from the high-resolution simulations discussed in Chap. 3, however, indicate an influence of the large-scale forcing on the shape and the width of the distribution. To analyze the impact of self-gravity on the density structure of turbulent gas, several methods can be applied. Clump finders search for extended gravitationally unstable regions. The resulting mass distribution, the so-called clump mass function, can be compared to analytical predictions. The employed stability criteria, however, are based on the Jeans or Bonnor-Ebert masses, which follow from linear perturbation analysis or the virial theorem for isolated objects. The dynamical equation for the rate of compression, on the other hand, applies to the fully non-linear regime of supersonic turbulence. By means of a statistical analysis of the relative contributions of gravitational, thermal, and turbulent source terms, the local support of the gas against gravity can be analyzed, without limitations imposed by analytical assumptions.

4.1 Global Averages and Probability Density Functions

As in Chap. 3, we begin our discussion with an illustrative case. The time evolution of the maxima and standard deviations of the mass density in the simulation ENZO_HYBR (see Table 3.1) shows that the largest overdensities are produced already after half an integral time scale T [1]. At this time, collisions of converging supersonic flows produce large sheetlike structures of shock-compressed gas, as can be seen in Fig. 4.1. These flows are a direct consequence of the mostly rotation-free forcing, which rarefies the gas in some regions and compresses the gas in between. At the vertices of the high-density sheets, cloud-like structures begin to form, which are highly

W. Schmidt, *Numerical Modelling of Astrophysical Turbulence*,
SpringerBriefs in Astronomy, DOI: 10.1007/978-3-319-01475-3_4,
© The Author(s) 2014

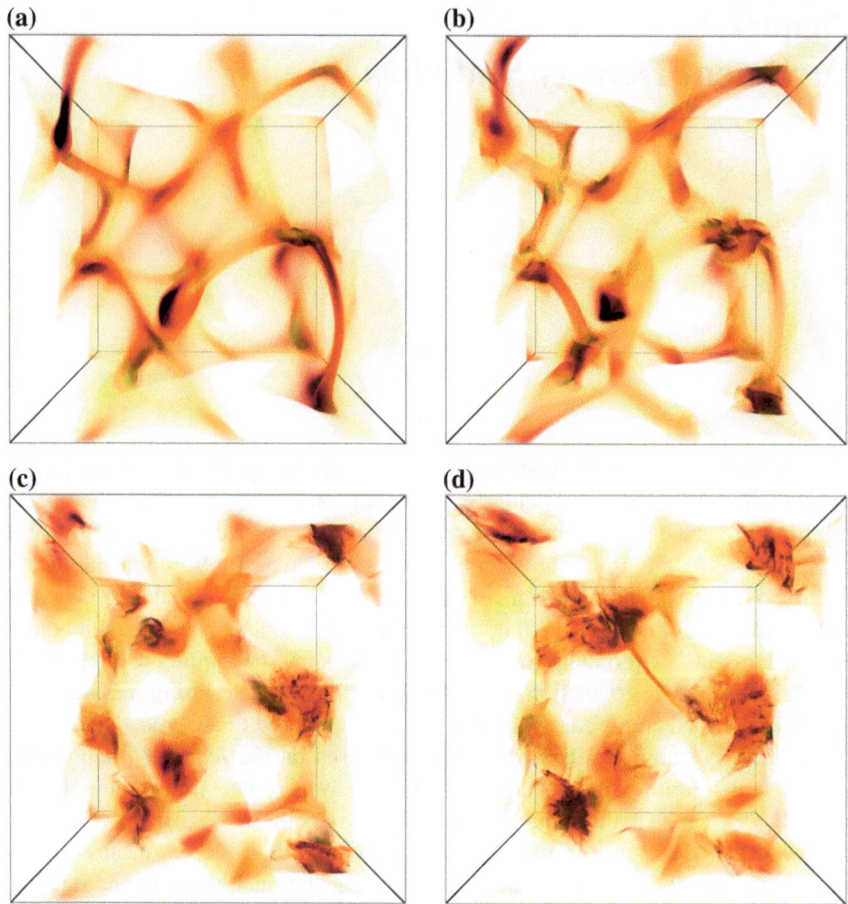

Fig. 4.1 Volume renderings of the mass density ρ for compressively driven turbulence with $\zeta = 0.1$ [1]. **a** $t = 0.44T$. **b** $t = 0.58T$. **c** $t = 0.73T$. **d** $t = 0.87T$

turbulent (compare to Fig. 3.1). Since the overdense gas is dispersed by this process, the peak densities decrease to values below 10^3.

The density statistics is specified by the probability density functions (PDFs) of the logarithmic density fluctuation s introduced in Sect. 1.2.2. The PDFs $p(s)$ at different instants are plotted in Fig. 4.2. The distribution quickly broadens and then settles into its final shape after about one integral time scale. The time-averaged PDF for $1 \leq t/T \leq 9$, which is over-plotted, is not symmetric but exhibits negative skewness, i. e., densities lower than the average are more probable than higher densities. This is suggested by the appearance of compact regions of over-dense gas surrounded by extended voids in Fig. 4.3. As a consequence, this PDF cannot be well fitted by the symmetric log-normal PDFs, which are defined by Eq. (1.32).

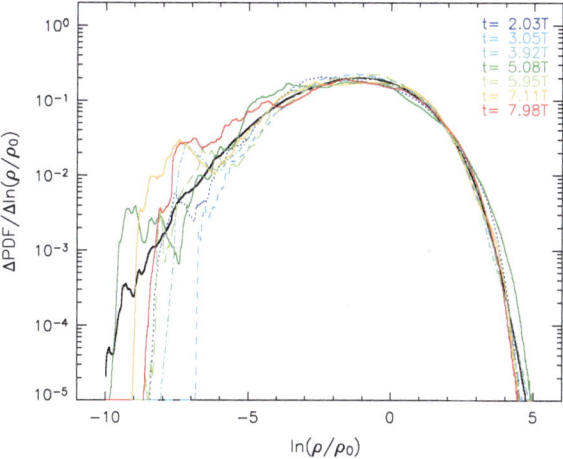

Fig. 4.2 Instantaneous and temporally averaged (thick *solid line*) probability density functions of the logarithmic mass density fluctuations (here, the acronym PDF in the vertical axis label signifies the probability distribution function, whose derivative is the probability density function) [1]

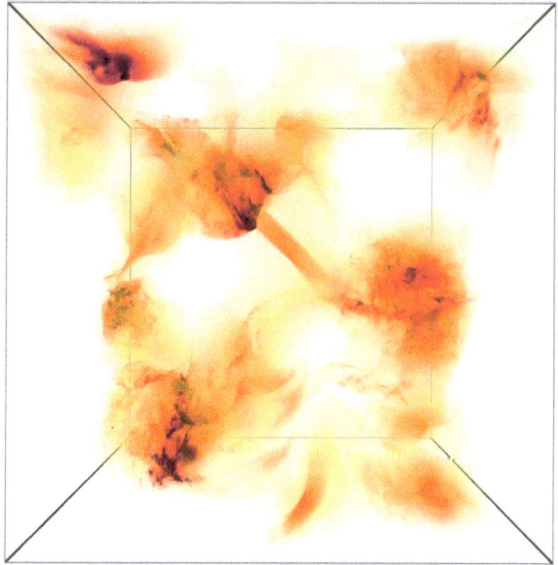

Fig. 4.3 Volume rendering of the mass density ρ after one integral time ($t = 1.02T$), continuing the evolution shown in Fig. 4.1

To understand why the theory of log-normal PDFs might break down, let us consider Eq. (1.71) for the rate of compression. For isothermal, non-gravitating gas, the equation simplifies to

$$-\frac{Dd}{Dt} = \frac{1}{2}\left(|S|^2 - \omega^2\right) + c_s^2 \nabla^2 s - \nabla \cdot \mathbf{f}.$$

In regions where $\nabla \cdot \mathbf{f} < 0$, the force tends to increase the rate of compression, i. e., it enhances compression if $d < 0$ and weakens rarefaction if $d > 0$. In the case $\nabla \cdot \mathbf{f} > 0$, the force has the opposite effect. For a stochastic force field, as defined by Eq. (2.7), positive and negative changes of d induced by the force field will occur with equal probability at any time at any position (by the very construction of the force field). For this reason, the forcing by itself does not induce an asymmetry. However, the divergence d is also affected by non-linear interactions. The PDF of d plotted in Fig. (2.6) indeed shows a distribution with strongly asymmetric tails, which implies a higher probability for rarefaction ($d > 0$) than for compression ($d < 0$) at any spatial position. For statistically stationary homogeneous turbulence, this also implies an asymmetry in the periods of time a particular fluid element is contracting or expanding. Equation (1.30) for the rate of change of the logarithmic density fluctuation s of a fluid element (the substantial time derivative of s) implies positive or negative increments $ds_\pm = -d_\mp \, dt$, depending on the sign of d. Since the probabilities for d_+ and d_- are not equal, the underlying assumption of a log-normal PDF of s (equal probabilities of ds_+ and ds_-) is violated.

The anomalous PDF that results from mainly compressive forcing is further investigated in [2, 3] for the two limiting cases of purely solenoidal and compressive forcing. As shown in Fig. 4.4, the PDF of the simulation FLASH_SOLN is relatively close to a log-normal distribution, but even in this case small deviations become apparent in the tails. A better fit is obtained with a skewed log-normal distribution [4]

$$p(s) = \frac{1}{\pi\,\omega}\,\exp\left[-\frac{(s-\xi)^2}{2\omega^2}\right]\int_{-\infty}^{(s-\xi)\alpha/\omega}\exp\left(-\frac{t^2}{2}\right)dt. \qquad (4.1)$$

The mean, the standard deviation, and the skewness of the distribution can be expresses in terms of the parameters α, ξ, and ω (see [2] for details). While a slightly negative skewness is found for solenoidal forcing, the PDF for purely compressive forcing (FLASH_COMP) has a pronounced skewness, and the tails are not well matched by analytical fit functions. These results indicate that rotation-free forcing, which directly modulates the divergence, affects the statistical independence of the density increments more than divergence-free forcing. In part, this could be a numerical resolution effect. As pointed out in [2], the density enhancements are limited by finite resolution. This means that density increments at very high densities cannot be statistically independent. Consequently, the far tails are numerically suppressed. Indeed, a dependence of the tails on numerical resolution can be seen in Fig. 4.5, but the changes are much smaller than the differences between solenoidal and compressive forcing.

Another important difference between solenoidal and compressive forcing is the much broader range of density fluctuations in the latter case. This is expected because compressive forcing directly induces density enhancements via the divergence coupling. A phenomenological relation between the width of a log-normal PDF, σ_s, and

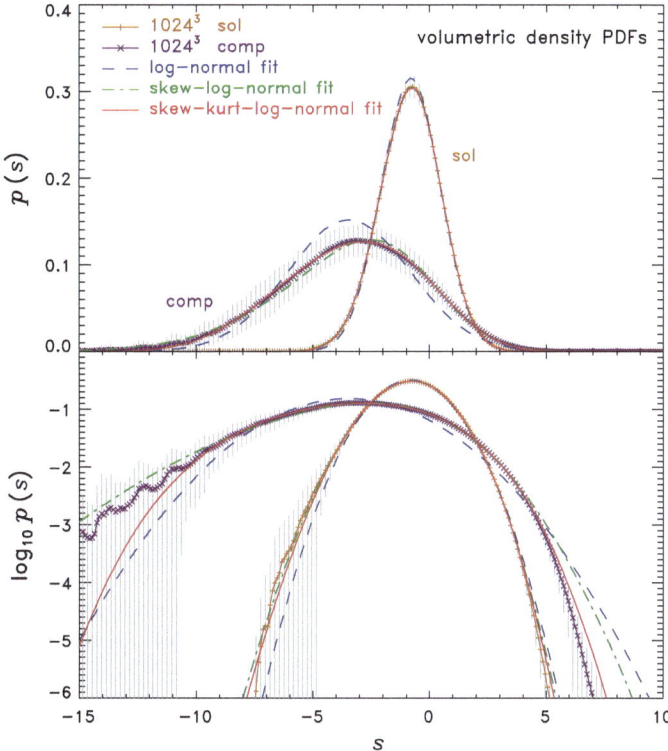

Fig. 4.4 PDFs of the logarithmic mass density fluctuations in simulations with purely solenoidal and compressive forcing [2]. The *top* plot with linear scale emphasizes the peaks, the logarithmic *bottom* plot the tails. Also shown are different analytical fit functions and, as *vertical lines*, the error *bars* of the numerical PDFs. By courtesy of Christoph Federrath

the RMS Mach number of turbulence is suggested in [5]:

$$\sigma_s^2 = \ln(1 + b^2 \mathcal{M}_{\rm rms}^2), \tag{4.2}$$

where $b \approx 0.5$ follows from MHD simulation data. Qualitatively, this relation simply states that the density fluctuations become stronger with increasing Mach number. For hydrodynamical turbulence with constant energy injection rate, $b \approx 0.26$ is reported in [6]. In [3], $b \approx 0.36$ is found for purely solenoidal forcing ($\zeta = 0$) and $b = 1.05$ for purely compressive forcing ($\zeta = 1$), while $b \approx 0.92$ for mainly compressive forcing with $\zeta = 0.1$ [1]. These findings can be explained by the $2 : 1$ ratio between transversal and longitudinal inertial-range modes if the forcing is solenoidal [3]. In this case, density fluctuations are solely induced by turbulent velocity fluctuations resulting from the non-linear energy transfer from larger scales. An isotropic compressive force field, on the other hand, randomly produces density fluctuations with equal probability in all spatial directions, resulting in a three times

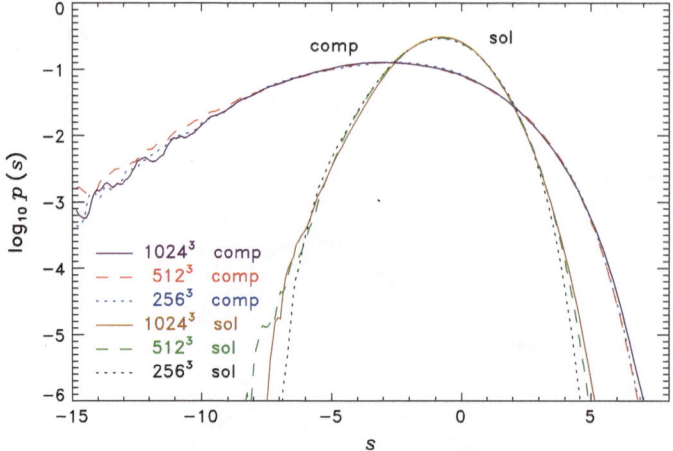

Fig. 4.5 PDFs as in Fig. 4.4 for three different numerical resolutions [2]. By courtesy of Christoph Federrath

stronger statistical weight. This conjecture is further investigated in [2] by computing the energy ratio of solenoidal and compressive modes for a series of simulations with different Helmholtz decomposition parameters ζ. As shown in Fig. 4.6, there is a correlation between this ratio and the coefficient b in Eq. (4.2), which is called the compressive ratio. For three-dimensional turbulence,

$$b \simeq \frac{\sqrt{3}\, E_{\text{long}}}{E_{\text{tot}}} = \frac{\sqrt{3}}{1 + E_{\text{trans}}/E_{\text{long}}}, \tag{4.3}$$

where E_{trans} and E_{long} are, respectively, the energies of transversal (solenoidal) and longitudinal (compressive) velocity fluctuations. Furthermore, a fit formula that relates b to ζ is proposed:

$$b(\zeta) \approx \frac{1}{D} + \frac{D-1}{D} \left[\frac{(1-\zeta)^2}{1 - 2\zeta + D\zeta^2} \right]^3. \tag{4.4}$$

The above phenomenological description is useful to estimate the influence of the forcing on the density statistics of supersonic turbulence. This is a potential input to star formation models [7] and to analytical theories of the clump mass function [8], for which the density PDF plays a central role (see Sect. 1.4.1).

Apart from the forcing, the randomness of density fluctuations is broken by self-gravity once the overdensity reaches the threshold to trigger gravitational collapse. By initializing a deep AMR simulation with a uniform-grid simulation of supersonic turbulence with $\mathcal{M}_{\text{rms}} \approx 6$, the evolution of self-gravitating turbulence is computed in [9]. The initial state is produced by a static driving force, composed of a spatially random mixture of solenoidal and compressive components. At $t = 0$, the forcing is

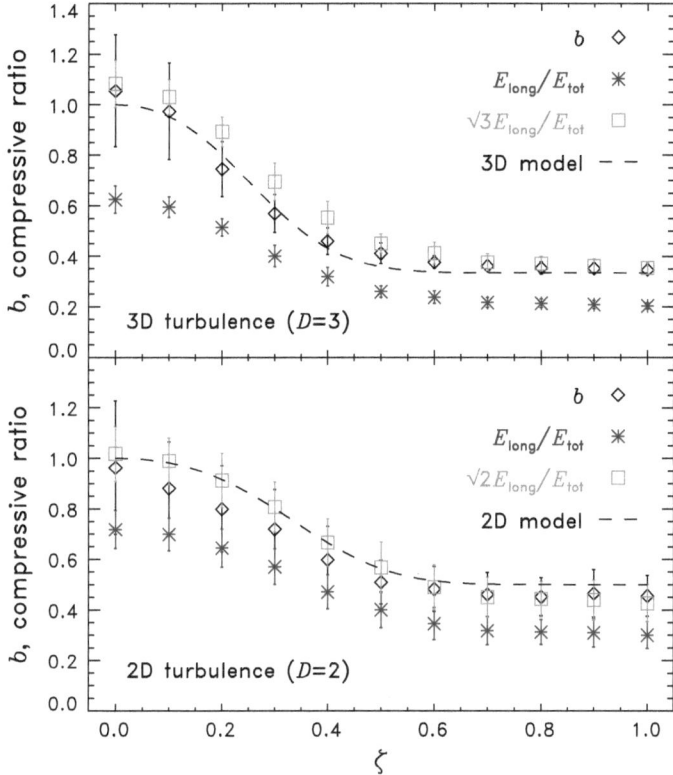

Fig. 4.6 The compressive ratio b following from Eq. (4.2) and ratio of the energy in longitudinal modes to the total energy in three- (*top*) and two-dimensional (*bottom*) simulations with different Helmholtz composition parameters ζ. The *dashed lines* show the fit functions defined by Eq. (4.4) [2]. By courtesy of Christoph Federrath

halted and self-gravity is activated. Subsequently, the collapse of the densest region is followed by five levels of AMR, with a refinement ratio of 4, such that the Jeans length given by the local density (Eq. 1.54 with ρ in place of ρ_0) is resolved by at least four grid cells [10]. Figure 4.7 shows PDFs of the density fluctuation ρ/ρ_0 at time $t = 0$, $0.26t_{\mathrm{ff}}$, and $0.42t_{\mathrm{ff}}$, where the typical time over which gravitational collapse proceeds is specified by the free-fall time scale,

$$t_{\mathrm{ff}}^{(0)} = \sqrt{\frac{3\pi}{32G\rho_0}} \approx 1.6\,\mathrm{Myr}\,, \tag{4.5}$$

for ρ_0 corresponding to a mean number density $n_0 = 500\,\mathrm{cm}^{-3}$. The gas temperature is 10 K, which is typical for molecular clouds, and the box size $X = 5$ pc. For these parameters, $X > \lambda_{\mathrm{J}}^{(0)}$, i. e., the gas in the box is globally unstable. Indeed, the

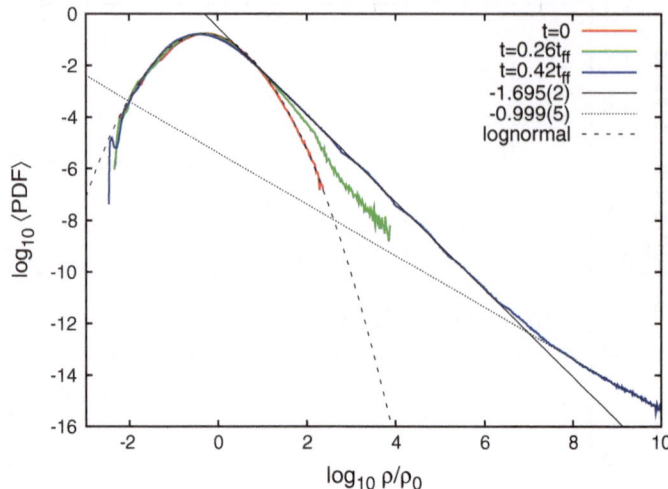

Fig. 4.7 PDF of the mass density fluctuation in an AMR simulation of self-gravitating turbulence at different instants [9]. The initial PDF is well approximated by a log-normal distribution. Also shown are power-law fits to sections of the final PDF at time $t = 0.42t_{ff}$, where t_{ff} is the free-fall time scale. By courtesy of Alexei Kritsuk

initially log-normal PDFs develops a power-law tail over a time of the order of the free-fall time scale. The slope of the tail is about -1.67. This can be interpreted as a signature of the self-similarity of collapsing structures. The flattening of the PDF at the highest densities corresponds to collapsed prestellar cores. However, these cores are not well resolved because the collapse is stopped by an artificial switch to adiabatic gas dynamics above the maximal resolvable density at the highest refinement level. In contrast to isothermal gas, adiabatic gas heats up through compression, which inhibits further collapse. Observations of star-forming clouds show PDFs with power-law tails, in agreement with simulation results, while log-normal PDFs are observed for molecular clouds in a state before star formation commences [11]. In a numerical parameter study encompassing a wide range of sonic and Alfvénic Mach numbers [12], it is further shown that the slope of the power-law tails in the self-gravitating regime flattens with increasing star formation efficiency, i. e., for higher mass fractions of collapsed cores.

4.2 Clump Mass Functions

For gravitationally unstable clumps produced by density fluctuations in supersonic turbulence, the theories outlined in Sect. 1.4.1 link the mass distribution of the clumps to the PDF of density fluctuations. This is a reasonable approximation as long as the PDF itself is not strongly influenced by self-gravity. Although collapsing cores eventually give rise to a power-law tail [9], one can still assume that the mass distribution of the clump-like structures that reach the critical mass for gravitational collapse is mainly determined by the turbulent density fluctuations. The consequences of this assumption are investigated by means of a clump-finding algorithm in [13, 14]. The algorithm divides the three-dimensional density field of supersonic isothermal turbulence into discrete density levels. Each density level is scanned for regions of connected cells with density values higher than the current density level. A connected region is identified as a clump candidate if its integrated mass exceeds the Bonnor-Ebert mass (see Sect. 1.4). In units of the solar mass M_{sun}, the Bonnor-Ebert mass is given by

$$m_{BE} = \frac{1.18 c_0^3}{G^{3/2} \rho^{1/2}} = 3.19 M_{sun} \left(\frac{\mu}{2.5}\right)^{-2} \left(\frac{\bar{n}}{1000\,cm^{-3}}\right)^{-1/2} \left(\frac{T}{10\,K}\right)^{3/2}, \quad (4.6)$$

where the gravitational constant $G = 6.67 \times 10^{-8}\,cm^3\,g^{-1}\,s^{-2}$, c_0 is the isothermal sound speed, T the temperature, and μ the mean molecular weight of the gas. The number density $n = \rho/(\mu m_H)$, where m_H is the mass of the hydrogen atom, is averaged over the connected region. Since the density increases toward higher levels, a clump candidate at a lower level can possibly be split into smaller subregions with masses greater than their Bonnor-Ebert masses at higher levels. In this way, clumps are recursively identified by the algorithm as connected regions that are gravitationally unstable and cannot be split further.

The clump finder was applied to the density fields from the FLASH_SOLN and FLASH_COMP simulations [14]. Since isothermal gas dynamics without self-gravity is invariant with respect to the the mean density ρ_0, which is obvious by considering Eqs. (1.30) and (1.31), an appropriate mass scale has to be chosen. The mass scale can be characterized by the total number N_{BE} of Bonnor-Ebert masses $m_{BE}(\rho_0)$ with respect to the mean density ρ_0 contained in the simulation box:

$$N_{BE} = \frac{m_{tot}}{m_{BE}(\rho_0)}$$
$$= 10^2 \left(\frac{\mu}{2.5}\right)^3 \left(\frac{n_0}{1000\,cm^{-3}}\right)^{3/2} \left(\frac{T}{10\,K}\right)^{-3/2} \left(\frac{2L}{1.782\,pc}\right)^3, \quad (4.7)$$

where $n_0 = \rho_0/(\mu m_H)$. The critical length scale corresponding to $m_{BE}(\rho_0)$ is the thermal Jeans length λ_J^0 with $a_J = 1.18$ (see Eq. 1.55), which can be expressed in units of the forcing length scale L as

$$\frac{\lambda_J^0}{L} = \frac{2L}{N_{BE}^{1/3}}. \quad (4.8)$$

The factor 2 stems from the ratio of the box size to L.

The mean number density chosen in [13] is $n_0 = 10^4\,cm^{-3}$, the domain size is $X = 6\,pc$, and $T = 10\,K$. In this case, $N_{BE} \approx 1.2 \times 10^5$ and $\lambda_J^0/L \approx 0.04$. The

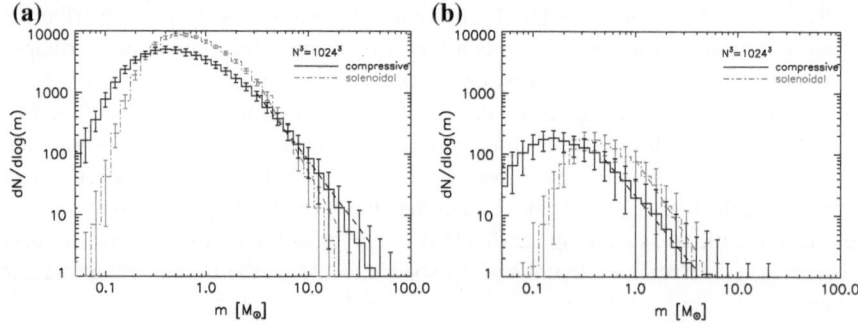

Fig. 4.8 CMFs for supersonic turbulence driven by solenoidal and compressive forcing [14]. The critical mass is given by the standard Bonnor-Ebert mass. The physical scales are chosen such that the following dimensionless parameters λ_J^0/L and N_{BE} are obtained. **a** $\lambda_J^0/L = 0.04$, $N_{BE} = 2 \times 10^5$. **b** $\lambda_J^0/L = 0.2$, $N_{BE} = 10^3$

Table 4.1 Least-square estimates of the power-law exponent x for the high-mass tails of the CMFs plotted in Fig. 4.8

λ_J^0/L	N_{BE}	x
Solenoidal forcing ($\zeta = 1.0$)		
0.04	1.2×10^5	3.1 ± 0.2
0.2	1.0×10^3	2.5 ± 0.7
Compressive forcing ($\zeta = 0.0$)		
0.04	1.2×10^5	2.1 ± 0.2
0.2	1.0×10^3	2.0 ± 0.4

CMFs calculated from the FLASH_SOLN and FLASH_COMP data for this set of parameters are plotted in Fig. 4.8a. Relative to the size Δ of the grid cells, $\lambda_J^0 \approx 20.8\Delta$. As demonstrated in [14], this entails the problem that many cores are poorly resolved with a linear extension over only a few cells. Another difficulty is that the above combination of density and size is atypical for observed molecular cloud properties. According to the Larson relation, $n_0 \approx 3000$ cm$^{-3}(L/1$ pc$)^{-1}$ [15, 16]. This suggests that the assumed forcing scale $L = X/2 = 3$ pc (see Sect. 2.1) is too large for molecular clouds of mean number density $n_0 = 10^4$ cm^{-3}. Without changing the temperature and the mean number density, consistency with the observed relation requires $L = 0.6$ pc, which implies $N_{BE} = 10^3$ and $\lambda_J^0/L \approx 0.2$. The resulting CMFs are also plotted in Fig. 4.8b. Although the total number of clumps is much smaller in this case, the dependence on resolution is significantly reduced, as shown in [14]. The slopes of least-square fits to the power-law tails of the CMFs are listed in Table 4.1. Regardless of the chosen mass scale, the values of x for both solenoidal and compressive forcing are significantly greater than the Salpeter value of 1.35, which is in qualitative agreement with the results for hydrodynamical turbulence in [13].

To compare the CMFs determined by the clump-finding algorithm with theoretical predictions, a semi-analytical approach is followed in [14]. As pointed out in Sect. 4.1,

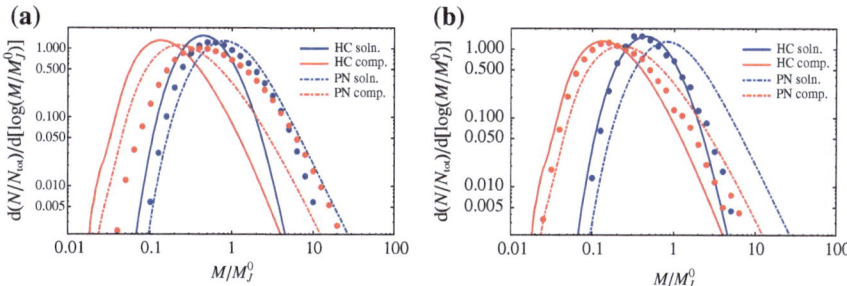

Fig. 4.9 Comparison of the normalized semi-analytic CMFs following from the Hennebelle-Chabrier (HC) theory and the Padoan-Nordlund (PN) theory to the corresponding mass distributions determined by a clump finder (*large dots*). The *thin dotted lines* are the tangents to the mass distributions with the Salpeter slope $x = 1.35$ [14]. **a** $\lambda_J^0/L = 0.04$, $N_{BE} = 2 \times 10^5$. **b** $\lambda_J^0/L = 0.2$, $N_{BE} = 10^3$

numerical PDFs of the mass density deviate from log-normal PDFs, particularly for compressive forcing. For this reason, the numerical data plotted in Fig. 4.4 are substituted into Eqs. (1.61) and (1.65). Since the total number of clumps depends on the chosen mass scale, the CMFs are normalized by N_{tot} (Eq. 1.58) for the comparison with semi-analytical CMFs. The results are shown in Fig. 4.9. Let us first consider the case $\lambda_J^0/L = 0.04$ ($N_{BE} = 2 \times 10^5$). For compressively driven turbulence, the clump-finding data match the theoretical predictions very poorly. Since the low-mass clumps cannot be resolved in this case, the distribution is biased towards higher masses in comparison to the semi-analytic distributions. For solenoidal forcing, the smallest clumps are at least marginally resolved, and the low-mass wing as well as the peak position agree reasonably well with the theoretical predictions. The high-mass wing, on the other hand, is only matched by the PN theory, although the slope of $x \approx 3.1$ (see Table 4.1) is significantly steeper than the theoretical value $x = 2.3$ following from Eq. (1.62) with $\beta = 1.86 \pm 0.05$ [2]. The opposite applies for $\lambda_J^0/L = 0.2$ ($N_{BE} = 10^3$). The numerical CMFs are close to the CMFs following from the Hennebelle-Chabrier theory, with deviations that are well within the error bars. In comparison to the Padoan-Nordlund theory, however, a large discrepancy of the high-mass tail becomes apparent for solenoidal forcing. Overall, the comparison suggests that Eq. (1.61) yields a good approximation only if λ_J^0/L is small and N_{BE} is large, which is an implicit assumption of the theory. Both the Padoan-Nordlund and the Hennebelle-Chabrier theory imply that the peak of the CMF is shifted towards lower masses for compressively driven turbulence, in agreement with the clump-finding data. This is a direct consequence of the shape of the mass-density PDFs. Since compressive forcing produces stronger density peaks, the fraction of low Jeans masses is higher compared to solenoidal forcing.

To account for the influence of the turbulent pressure on the support of the gas against gravity, the stability criterion can be modified by replacing the thermal speed of sound in the definition of the Bonnor-Ebert mass with an effective speed of sound,

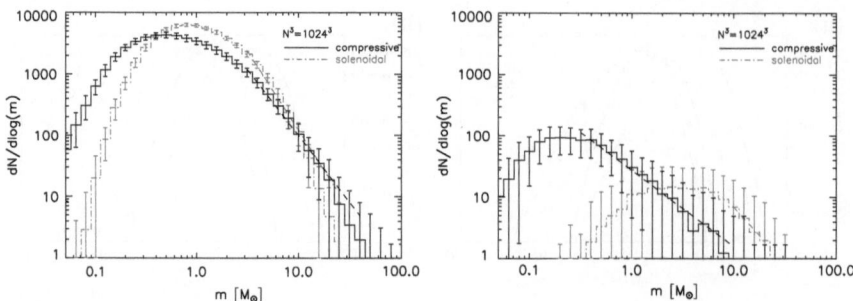

Fig. 4.10 CMFs as in Fig. 4.8, but with the turbulent velocity dispersion of the clumps incorporated into the stability criterion [14]

Table 4.2 Least-square estimates of the power-law exponent x for the high-mass tails of the CMFs plotted in Fig. 4.10

Λ_J^0/L	N_{BE}	x_{LS}
Solenoidal forcing, $\zeta = 1.0$		
0.04	1.2×10^5	2.8 ± 0.2
0.2	1.0×10^3	-
Compressive forcing, $\zeta = 0.0$		
0.04	1.2×10^5	2.1 ± 0.2
0.2	1.0×10^3	1.2 ± 0.3

as proposed in [17]:

$$c_{\text{eff}}^2 = c_0^2 + \frac{1}{3}\sigma_{\text{clump}}^2. \tag{4.9}$$

The turbulent velocity dispersion σ_{clump} of a clump is computed from the mass-weighted RMS velocity fluctuation with respect to the centre-of-mass velocity for all grid cells that belong to the clump. The mass distributions of clumps supported by the effective pressure are plotted in Fig. 4.10 for the two different mass scales. While the turbulent pressure has no significant impact for $N_{BE} = 1.2 \times 10^5$, it clearly changes the CMFs for $N_{BE} = 10^3$. In the case of solenoidal forcing, turbulent support leads to a large shift of the peak. This is a consequence of the significantly larger cores for $N_{BE} = 10^3$. In the compressive case, the peak of the CMF is less affected, but a flattening of the high-mass tail from $x \approx 2.0$ to $x \approx 1.2$ can be observed (see Table 4.2), which is close to the Salpeter slope.

The CMFs with the stability criterion based on Eq. (4.9) can be compared with the semi-analytical CMFs given by Eqs. (1.66) and (1.68), as detailed in [14]. For $\lambda_J^0/L = 0.04$, Fig. 4.11 shows that there is only a weak effect of turbulent pressure in the case of compressive forcing because the velocity fluctuations are too small on the clump scale R and the clumps remain under-resolved. However, a significantly improved agreement with the semi-analytical CMF is obtained for solenoidal forcing. In this case, the clumps tend to be larger and turbulent pressure becomes significant relative

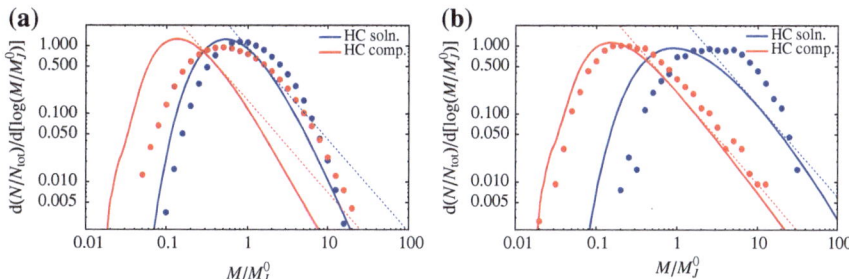

Fig. 4.11 Comparison of the normalized semi-analytic CMFs following from the Hennebelle-Chabrier (HC) theory with effective pressure support to the corresponding mass distributions determined by a clump finder (*large dots*). The *thin dotted lines* are the tangents to the mass distributions with the Salpeter slope $x = 1.35$ [14]. **a** $\lambda_J^0/L = 0.04$, $N_{\mathrm{BE}} = 2 \times 10^5$. **b** $\lambda_J^0/L = 0.2$, $N_{\mathrm{BE}} = 10^3$

to the thermal pressure. The same applies to the CMF produced by compressive forcing if $\lambda_J^0/L = 0.2$. While Eq. (1.62) yields $x \approx 2.7$ for the spectral index $\beta = 1.94 \pm 0.05$ of compressively driven turbulence [2], the Hennebelle-Charbier theory is in very good agreement with the much flatter power-law tail resulting from the clump-finding analysis with effective pressure support. This suggests that the Hennebelle-Chabrier theory reproduces the properties of larger clumps well if the effect of turbulent pressure is incorporated into the critical mass. For solenoidal forcing, however, the detailed analysis in [14] shows that the length scale of the larger clumps approaches the integral scale L if $\lambda_J^0/L = 0.2$. In this case, the approximate Eq. (1.68) breaks down. Moreover, it is not at all clear that the simple Bonnor-Ebert mass, which strictly applies to spherical objects of uniform density, correctly captures the stability properties of extend clumps in a highly turbulent medium, particularly if the clumps possess a complicated internal structure. Consequently, also the clump-finding results could be significantly biased.

4.3 Local Support Against Gravity

According to the Jeans criterion, the stability against gravitation collapse increases with the thermal pressure of the gas (Eq. 1.54). On the other hand, the local thermal support Λ_{therm} (Eq. 1.74) depends on the pressure Laplacian and the density and pressure gradients. Is there a connection between these two notions of support? In the following, data from the AMR simulation of self-gravitating turbulence introduced in Sect. 4.1 are used to investigate this question [9, 18].

A quantity with the physical dimension of pressure can be obtained by multiplying Λ_{therm} with the gas density ρ and the square of the local grid scale Δ (numerical differentiation corresponds to the division of pressure differences between adjacent cells by Δ). The resulting quantity $\Delta^2 \rho \Lambda_{\mathrm{therm}}$, however, can be positive or negative. Depending on the sign, Λ_{therm} counteracts or enhances gravitational compression. To calculate statistics of Λ_{therm}, it is convenient to separate the local support into

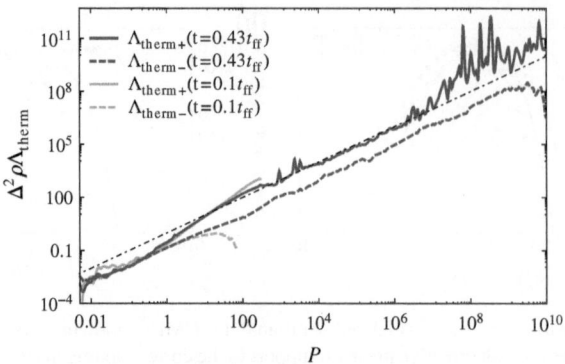

Fig. 4.12 Averages of the positive and negative thermal support $\Delta^2 \rho \Lambda_{\text{therm}\pm}$ as function of the thermal pressure at time $t = 0.43t_{\text{ff}}$ in code units (normalization to unit mean density). The positive and negative components are shown as *solid* and *dashed lines* and the identity function is indicated by a *dot-dashed line*. For comparison, the light-colored lines show the support functions for $t = 0.1t_{\text{ff}}$ [18]

positive and negative components, which are generically defined by

$$\Lambda_+ = \begin{cases} \Lambda & \text{if } \Lambda \geq 0, \\ 0 & \text{otherwise,} \end{cases} \qquad \Lambda_- = \begin{cases} -\Lambda & \text{if } \Lambda \leq 0, \\ 0 & \text{otherwise.} \end{cases} \qquad (4.10)$$

The mean positive and negative components add up to the mean net support, i. e.,

$$\langle \Lambda \rangle = \langle \Lambda_+ \rangle - \langle \Lambda_- \rangle$$

both for volume- and mass-weighted averaging. If the net effect of thermal pressure is to support the gas, as is expected from the Jeans analysis (see Sect. 1.4.2), then $\langle \Lambda_{\text{therm}} \rangle$ should be positive.

The volume-weighted mean values of $\Delta^2 \rho \Lambda_{\text{therm}\pm}$ plotted in Fig. 4.12 show that indeed the positive component of the thermal support dominates for high pressure.[1] Since gravitational collapse produces much higher densities than the initial supersonic turbulence, the range of pressures also increases greatly over time. This can be seen by comparing to the early instant $t = 0.1t_{\text{ff}}$, where the free-fall time scale t_{ff} is defined by Eq. (4.5). For $t = 0.43t_{\text{ff}}$, the pressure-averaged values can be approximated by the asymptotic relation

$$\langle \Delta^2 \rho \Lambda_{\text{therm}} \rangle_P \simeq P \quad \text{for } P \gtrsim 100.$$

A value of unity corresponds to the thermal pressure at the mean density. In this sense, a higher pressure corresponds to an enhanced support against gravity, as expressed by the formula for the Jeans length. But even for very high pressures, there is a non-vanishing fraction of negative support. Since $P \simeq c_0^2 \rho$, the above relation implies

---

[1] Numerically, $\Delta^2 \rho \Lambda_{\text{therm}\pm}$ is averaged over narrow bins of P.

the typical magnitude $\Lambda_{\text{therm}} \sim (c_0/\Delta)^2$ at high densities. Naturally, the support functions depend on the numerical grid resolution Δ because they are determined by derivatives. Physically, the derivatives are limited by the viscous dissipation scale and the length scale on which the collapse of the gas stops. For a discussion of the dependence on numerical resolution, see [18]. The strong fluctuations for pressures higher than about 10^7 are due the increasinlgy small samples, which are strongly affected by extreme local conditions.

To analyze the influence of turbulence on the support of the gas, we consider $\rho\Delta^2\Lambda_{\text{turb}}$, where Λ_{turb} is defined by Eq. (1.73). The first term, $\frac{1}{2}\Delta^2\rho\omega^2 = \Delta^2\Omega$, where $\Omega = \frac{1}{2}\rho\omega^2$ is the enstrophy density, can be interpreted as the turbulent pressure due to eddies on the grid scale Δ. The second term, $-\frac{1}{2}\Delta^2\rho|S|^2$, corresponds to a negative pressure that is produced by the strain of the flow. Since the trace of the rate-of-strain tensor is the divergence d, the strain becomes particularly large in the vicinity of shocks. Averages of the positive and negative turbulent support,

$$\Delta^2\rho\Lambda_{\text{turb}+} = \begin{cases} \Delta^2\rho(\omega^2 - |S|^2)/2 & \text{if } \omega > |S|, \\ 0 & \text{otherwise}, \end{cases}$$

$$\Delta^2\rho\Lambda_{\text{turb}-} = \begin{cases} \Delta^2\rho(|S|^2 - \omega^2)/2 & \text{if } \omega < |S|, \\ 0 & \text{otherwise}, \end{cases}$$

are plotted as functions of $\Delta^2\Omega$ in Fig. 4.13. Remarkably, the negative component $\Delta^2\rho\Lambda_{\text{turb}-}$ is greater than $\Delta^2\rho\Lambda_{\text{turb}+}$ for all enstrophy densities. This implies that the local turbulent support is dominated by shock compression, while the positive support caused by turbulent eddies is subdominant. Except for very low enstrophy, both components of the turbulent support closely follow linear relations and the total turbulent support function is approximately given by the *negative* turbulent pressure associated with $\Delta^2\Omega$:

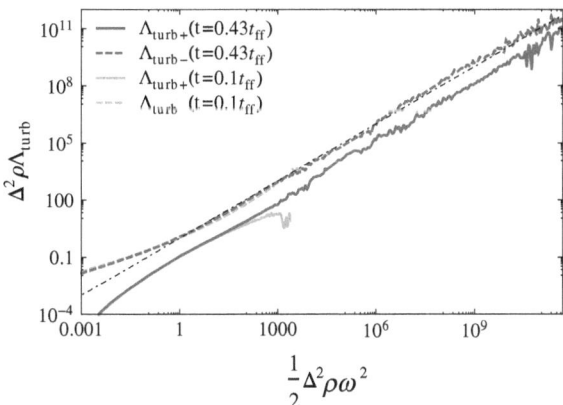

Fig. 4.13 Averages of the positive and negative turbulent support $\Delta^2\rho\Lambda_{\text{turb}\pm}$ versus enstrophy density at time $t = 0.1t_{\text{ff}}$ and $0.43t_{\text{ff}}$ [18]

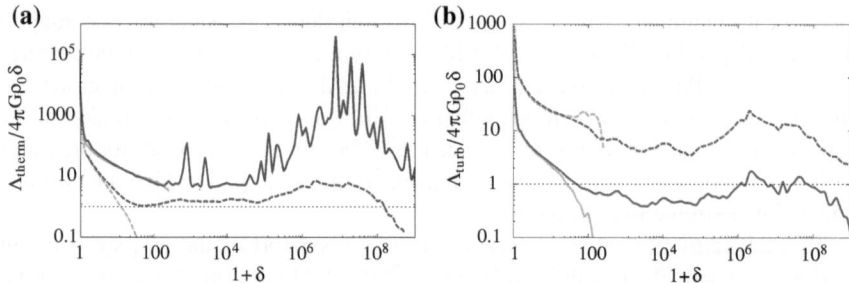

Fig. 4.14 Averages of the thermal (**a**) and turbulent (**b**) support relative to the gravitational compression rate as functions of the overdensity at time $t = 0.1t_{ff}$ (*light color*) and $t = 0.43t_{ff}$ (*full color*) [18]. As in Figs. 4.12 and 4.13, positive and negative components are shown as *solid* and *dashed lines*

$$\langle \Delta^2 \rho \Lambda_{turb} \rangle_{\Delta^2 \Omega} \simeq -\Delta^2 \Omega \quad \text{if } \Delta^2 \Omega \gtrsim 1. \tag{4.11}$$

The scale factor Δ^2 in this relation does not cancel out here because values from different refinement levels contribute to the averages.

In Fig. 4.14, the ratios of Λ_{therm} and Λ_{turb} to the gravitational compression rate $4\pi G\rho_0\delta$ rate are plotted as functions of the overdensity $1 + \delta = \rho/\rho_0$. In the early phase ($t = 0.1t_{ff}$), the the support functions simply decrease with density. For $t = 0.43t_{ff}$, the statistics plotted in Fig. 4.14 clearly show that the positive component of the thermal support and the negative component of the turbulent support dominate for all $\delta > 0$. The total support relative to the gravitational compression rate, $\Lambda/(4\pi G\rho_0\delta)$, is plotted in Fig. 4.15. At time $t = 0.1t_{ff}$, the critical value of unity is approached for the highest densities ($\delta \sim 100$), which indicates that the gas is self-gravitating. Since $\Lambda_{turb}-/4\pi G\rho_0\delta$ is large for almost all densities, the gas is mainly compressed by shocks. Compared to $t = 0.1t_{ff}$, the support at time $t = 0.43t_{ff}$ remains nearly unaltered for the lower range of densities, but an intermediate range $100 \lesssim \delta \lesssim 10^5$ has formed, in which the ratio of support to gravity is nearly constant and $\Lambda_{therm}+$ is roughly balanced by $\Lambda_{turb}-$. Except for the strong spikes, which probably correspond to collapsing filaments that start to feel strong pressure support, the net support is negative and of the order of the gravity term (the median value is $\Lambda/4\pi G\rho_0\delta \approx -0.67$; see [18]). This can be interpreted as the onset of gravitational collapse, triggered by supersonic gas compression. At densities around 10^7, however, the thermal pressure of the gas tends to strongly overcompensate gravity. This does not necessarily mean that the gas is expanding ($d > 0$), but that the contraction slows down ($Dd/Dt > 0$) toward higher densities. For $\delta \gtrsim 10^7$, both the thermal and the turbulent support decrease relative to the gravitational compression rate. In [9], it is argued that overdensities $\sim 10^7$ mark the transition to rotationally supported cores, which entails a decreasing importance of thermal support. This transition can be seen as a flattening of the power-law slope of the density PDF shown in Fig. 4.7. For the highest densities, however, the gas dynamics is not well resolved. Samples of dense disk-like cores together with PDFs of the gas density in cubic regions centered

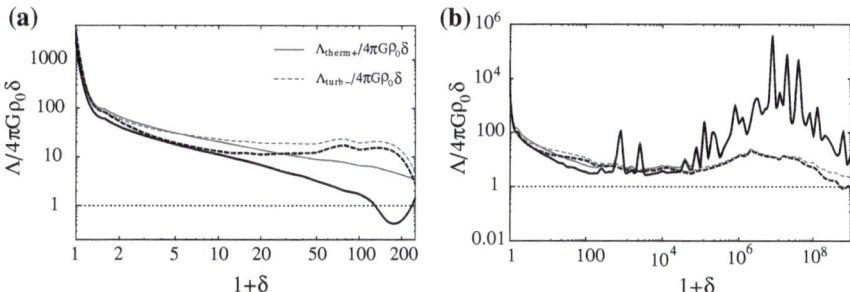

Fig. 4.15 Plots of the positive (*thick solid*) and negative (*thick dashed*) components of the total support at time $t = 0.1 t_{\text{ff}}$ (**a**) and $t = 0.43 t_{\text{ff}}$ (**b**). Also shown are the dominant components of the thermal and turbulent support functions (*thin lines*) [18]

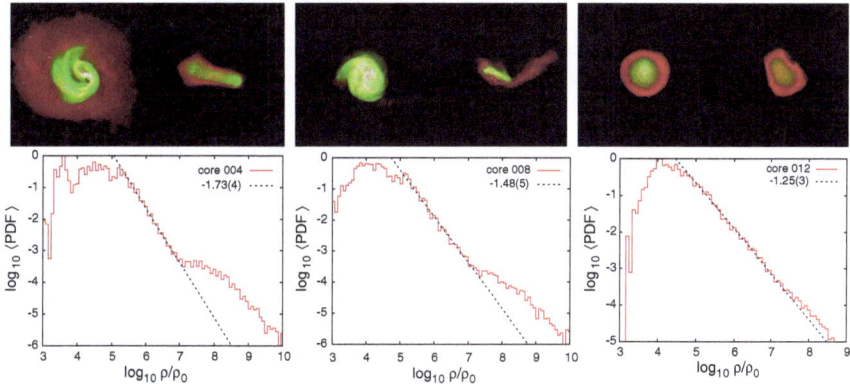

Fig. 4.16 Volume rendering (*top* three panels) and PDFs (*bottom*) of the mass density for three extracted cores with maximum densities (from *left* to *right*) $\rho_{\text{max}}/\rho_0 = 8.5 \times 10^{10}, 3.2 \times 10^{10}$, and 6.0×10^8 [9]. By courtesy of Alexei Kritsuk

at the density peaks are shown in the left and middle panels of Fig. 4.16, while the right panel shows a more globular core at lower density.

The lack of a positive turbulent support is contrary to the heuristic formula (1.57), which is applied to the statistical analysis of gravitationally unstable clumps in Sect. 4.2. In principle, positive turbulent support could occur for extended clumps, but locally shock compression appears to be the dominant effect. The simulations of self-gravitating turbulence presented in [19] support the conclusion that the main role of turbulence is to compress the gas to critical densities at which gravitational collapse commences. These simulations use the so-called sink particle technique. This means that gas from collapsing regions is removed from the hydrodynamical density field and absorbed by numerical particles. The mass of the particles, which roughly correspond to pre-stellar cores, can be interpreted as collapsed gas mass. The conversion of gas into sink particles is controlled by the the criteria specified

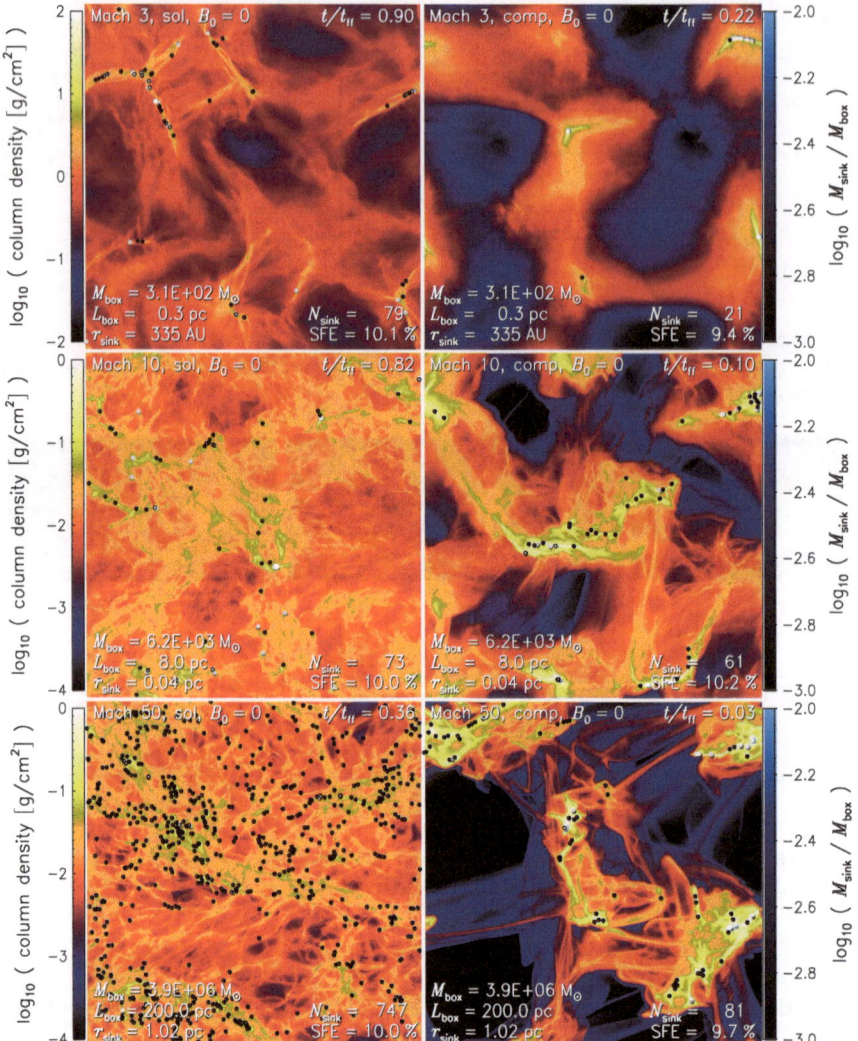

Fig. 4.17 Column density projections from simulations of self-gravitating turbulence with solenoidal forcing (*left* panels) and compressive forcing (*right* panels) for Mach numbers 3 (*top*), 10 (*middle*), and 50 (*bottom*). Collapsing gas is captured by sink particles, which are indicated by the *small dots*. Each snapshot corresponds to the time when 10 % of the total mass in the simulation domain are converted into sink particles [19]. By courtesy of Christoph Federrath

in [20]. In Fig. 4.17, the projected density fields from six simulations are shown. The projection sums up the density values in one spatial direction to obtain column densities. Solenoidal and compressive forcing is applied, with steady-state Mach numbers ranging from 3 to 50. All snapshots shown in Fig. 4.17 have a total collapsed mass of about 10 % of the initial gas mass in the box, but this mass fraction is reached earlier for higher Mach numbers, as indicated by the simulation time in

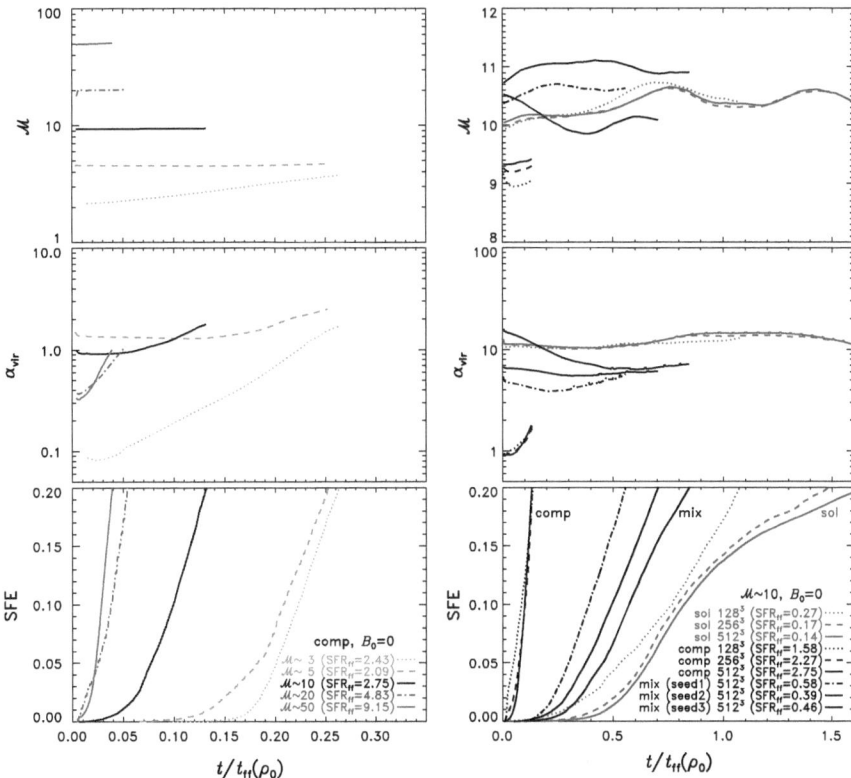

Fig. 4.18 Time evolution of the RMS Mach number (*top*), virial parameter (*middle*), and star formation efficiency (*bottom*) for compressively driven turbulence at different Mach numbers (*left*) and different mixtures of solenoidal and compressive forcing at Mach 10 (*right*) [19]. By courtesy of Christoph Federrath

units of $t_{\rm ff}$. One can also see that the number of sink particles increases with the Mach number. For a given Mach number, gravitational collapse proceeds faster for compressive forcing. The sink particles also tend to be more clustered, corresponding to the different distribution of the gas in the compressive case (see Sect. 4.1).

Furthermore, the enhancement of gravitational collapse by turbulence can be deduced from a plot of the mass fraction of sink particles as a function of time for various Mach numbers (see left plot in Fig. 4.18). The time-dependent collapsed mass fraction is interpreted as star formation efficiency (SFE). Also shown is the time evolution of the RMS Mach number (simulations with faster production of sink particles are stopped earlier) and the virial parameter $\alpha_{\rm vir} = 2E_{\rm kin}/|E_{\rm grav}|$, where $E_{\rm kin}$ and $|E_{\rm grav}|$ are, respectively, the total kinetic and gravitational energies. For homogeneous isotropic turbulence, $E_{\rm kin}$ corresponds to the turbulent energy. The typical value of the virial parameter in the simulations is around unity, which indicates an equipartition between the global turbulent and gravitational energies.

In the right plot, the SFE is plotted for solenoidal ($\zeta = 1.0$), mixed ($\zeta = 0.5$), and compressive ($\zeta = 0.0$) forcing for different numerical resolutions and random seeds of the Ornstein-Uhlenbeck processes (see Sect. 2.1). For purely compressive forcing, the rate of sink particle production is greater by an order of magnitude than for solenoidal forcing. Generally, the dependence on the forcing is substantially more pronounced than the influence of the numerical resolution. The random seed has a stronger influence, but is still subdominant. The turbulent energy E_{kin} is nearly the same for all simulations, but $|E_{grav}|$ tends to be higher for compressive forcing because of the stronger density enhancements. Consequently, α_{vir} tends to be higher for solenoidal forcing, as gravity is weaker relative to turbulence. The increase of the SFE for higher Mach number and larger fraction of compressive modes indicates that stronger supersonic compression produces a larger amount of collapsed gas, which qualitatively agrees with the results for the statistics of the local turbulent support Λ_{turb}.

References

1. W. Schmidt, C. Federrath, M. Hupp, S. Kern, J.C. Niemeyer, A&A **494**, 127 (2009). doi:10.1051/0004-6361:200809967
2. C. Federrath, J. Roman-Duval, R.S. Klessen, W. Schmidt, M. Mac Low, A&A **512**, A81+ (2010). doi:10.1051/0004-6361/200912437.
3. C. Federrath, R.S. Klessen, W. Schmidt, ApJ **688**, L79 (2008). doi:10.1086/595280
4. A. Azzalini, Scand. J. Statist. **12**, 171 (1985)
5. P. Padoan, A. Nordlund, B.J.T. Jones, MNRAS **288**, 145 (1997)
6. A.G. Kritsuk, M.L. Norman, P. Padoan, R. Wagner, ApJ **665**, 416 (2007). doi:10.1086/519443
7. H. Braun, W. Schmidt, MNRAS **421**, 1838 (2012). doi:10.1111/j.1365-2966.2011.19889.x
8. P. Hennebelle, G. Chabrier, ApJ **702**, 1428 (2009). doi:10.1088/0004-637X/702/2/1428
9. A.G. Kritsuk, M.L. Norman, R. Wagner, ApJ **727**, L20 (2011). doi:10.1088/2041-8205/727/1/L20
10. J.K. Truelove, R.I. Klein, C.F. McKee, J.H. Holliman II, L.H. Howell, J.A. Greenough, ApJ **489**, L179 (1997). doi:10.1086/316779
11. J. Kainulainen, H. Beuther, T. Henning, R. Plume, A&A **508**, L35 (2009). doi:10.1051/0004-6361/200913605
12. C. Federrath, R.S. Klessen, ApJ **763**, 51 (2013). doi:10.1088/0004-637X/763/1/51
13. P. Padoan, Å. Nordlund, A.G. Kritsuk, M.L. Norman, P.S. Li, ApJ **661**, 972 (2007). doi:10.1086/516623
14. W. Schmidt, S.A.W. Kern, C. Federrath, R.S. Klessen, A&A **516**, A25 (2010). doi:10.1051/0004-6361/200913904
15. R.B. Larson, MNRAS **194**, 809 (1981)
16. E. Falgarone, J. Puget, M. Perault, A&A **257**, 715 (1992)
17. S. Chandrasekhar, Proc. Roy. Soc. London Ser. A **210**, 26 (1951)
18. W. Schmidt, D.C. Collins, A.G. Kritsuk, MNRAS (2013). doi:10.1093/mnras/stt399
19. C. Federrath, R.S. Klessen, ApJ **761**, 156 (2012). doi:10.1088/0004-637X/761/2/156
20. C. Federrath, R. Banerjee, P.C. Clark, R.S. Klessen, ApJ **713**, 269 (2010). doi:10.1088/0004-637X/713/1/269

Index

W. Schmidt, *Numerical Modelling of Astrophysical Turbulence,*
SpringerBriefs in Astronomy, DOI: 10.1007/978-3-319-01475-3,
© The Author(s) 2014